T0255061

A Full Axiomatic Development of High School Geometry

David M. Clark • Samrat Pathania

A Full Axiomatic Development of High School Geometry

David M. Clark
New Paltz, NY, USA

Samrat Pathania
Kerhonkson, NY, USA

ISBN 978-3-031-23527-6 ISBN 978-3-031-23525-2 (eBook)
https://doi.org/10.1007/978-3-031-23525-2

Cover illustration: Cover image used by permission of David Clark and the American Mathematical Society.

This Springer imprint is published by the registered company Springer Nature Switzerland AG
The registered company address is: Gewerbestrasse 11, 6330 Cham, Switzerland

This book is dedicated to

Annabella, Tommy

and all young scholars of mathematics.

Preface

This textbook is written for those who seek a full understanding of the topics that form the standard content of high school Euclidean geometry. After giving a complete axiomatic development of those topics, it concludes with a proof of the consistency of the axioms used and a full description of their models. It is given in guided inquiry (inquiry-based) format with the intention that students will prove the theorems and present their proofs to their class with the instructor as a mentor and a guide.

The book is written for graduate and advanced undergraduate students interested in teaching secondary school mathematics, for pure math majors interested in learning about the foundations of geometry and for college and university faculty preparing future secondary school teachers. Because the presentation is fully self contained, it is suitable for self-study by advanced math students and professional mathematicians. It constitutes the top tier of a sequence of four educational tiers of geometry, beginning at the preschool level. We offer it with the hope that it will help to bring that sequence together into a coordinated and integrated whole.

At the first tier, young children gain experiences with round objects, like balls, wheels, marbles and oranges. They also get experience with rectilinear objects like building blocks, books and houses and linear objects like sticks and strings. Soon words for abstract shapes and forms, like "circle", "square" and "line", help to solidify those categories. Eventually those words get precise definitions as they discover that some words have an exact meaning that everyone can understand.

The second tier is high school, where geometry becomes a real subject of its own. There students are given an assortment of facts of geometry, sometimes called *axioms*, sometimes *postulates* and sometimes *theorems*, that they are free to use. Their task is to understand these facts and the associated definitions and then use them to solve problems and establish other facts. Completing these tasks requires making logical arguments, orally or in writing, and learning to justify those arguments.

Everyone needs to be able to reason from evidence, to draw valid conclusions and to justify those conclusions to others on a regular basis. A functioning democracy requires voting citizens who can make sound judgements about the validity of the arguments they hear. In short, we all have a stake in having these thinking, reasoning and communicating skills passed down in our schools to the next generation. Teachers of all subjects at all levels have an opportunity to help do this, and many do so very well.

But there is one place in the standard high school curriculum that offers an opportunity to put a particularly direct focus on these skills. That is *Euclidean geometry.* Solving problems by applying definitions and given information allows geometry students to draw conclusions that they can justify and defend with a kind of authority that in no way depends on their age, gender, race or socio-economic status. It empowers them with skills that can be invoked in any pursuit that they might subsequently follow.

At the third tier teachers need to get preparation to convey these benefits to high school students. Normally this must come from a one semester undergraduate geometry course. In order that they can teach the high school geometry course their students will need, it is our view that a college geometry course for teachers should fulfill three goals.

Structure: Mathematical Proof. A college level study of a subject is not just about what is true in that particular subject, but also about how truth is established in that subject. In the context of geometry, this means that teachers need to understand the source and genesis of the facts they will be giving their own students. As in all of mathematics, these facts are established by means of a mathematical proof.

Pedagogy: Guided Inquiry. A geometry course for pre-service teachers should offer them a direct experience with the kind of active learning pedagogy we would like them to use later with their own students. At the college level, this means that *they* will solve the problems, prove the theorems, and present their results to their classmates with the instructor as a mentor and a guide.

Content: High School Geometry. Geometry is a beautiful subject offering a vast array of interesting topics that could be studied. Pre-service teachers need a course that treats exactly those topics that they will need to teach in the schools. In spite of the many variations we have seen in high school geometry, that list of topics has remained remarkably stable over time.

Most college level geometry courses have a focus on mathematical proof. But many are still not using a guided inquiry (inquiry-based or active learning) pedagogy. Yet there is now an extensive body of research demonstrating the benefits of active learning. This is particularly applicable to the learning of mathematical proof. Students cannot learn to prove theorems by watching someone else prove theorems any more than they can learn to play the piano by listening to someone else play the piano. Similarly few college level geometry courses focus exclusively on the topics that are taught in high school.

There are many other interesting topics of geometry that can be valid components of a college level course. But it is our view that future high school teachers need first and foremost, in the one geometry course they are likely to have, to gain a deep understanding of exactly those topics they will need to teach.

In order to prove the important facts of high school geometry with a guided inquiry format in a single undergraduate semester, it is important to look closely at the nature of mathematical proof that is appropriate in this context. It is simply unfeasible to prove every detail using a set of rudimentary axioms like those used by David Hilbert [8]. Not only is it unfeasible. It is not what future teachers are ready for at this stage. What they need is a course that takes a relaxed view of unstated assumptions, as was always used in teaching Euclid's geometry. They then need to be given selected major theorems of geometry as axioms, and then to prove consequences of those axioms. This format will match the teaching they will need to do, but will involve more theorem proving and less problem solving.

We are aware of only one third tier undergraduate geometry text that achieves all three of these goals. That is Clark's *Euclidean Geometry: A Guided Inquiry Approach* [2]. That book looks at a plane as a set of points with length, area and angle measures as primitive constructs governed by axioms listing each of their properties. Five other axioms give properties of congruence, parallels and similarity. Students then prove the other important facts of geometry using these axioms and a limited range of unstated assumptions.

It is our view that, in order to secure the reliability and credibility of K-12 geometry, this enterprise needs to maintain within its academic leadership a critical number who have a full and complete understanding of the subject. By this we mean that they fully understand the source and genesis of the content of K-12 geometry with no unstated assumptions or other logical gaps or compromises. It was with this conviction that we set out to write this book, offering a fourth tier textbook for serious students of mathematics wanting to focus on K-12 education. To do this we look at a plane as a set of points with a distance function on pairs of points as its only primitive construct. Simple axioms in Chapter 1 are used to prove the commonly unstated assumptions of plane geometry. From there the notions of length, area and angle measure, along with congruence and similarity, are carefully defined and their properties proven as theorems. For the benefit of those familiar with [2], we have added Appendix B where we list all of the mathematical and logical gaps in [2] and outline where and how they have been filled in this text.

Using this book to fill those gaps will require a good bit of mathematical maturity. In contrast with [2] we are starting with simpler, more rudimentary axioms and therefore need to do more work to get the same results. For example, many geometry texts, and particularly high school texts, define segments as congruent if they have the same length measure, angles as congruent if they have the same degree measure and triangles as congruent if they have congruent sides and angles. But the notion of congruence should really apply to all figures. Circles with the same radius should count as congruent just as the copies of page x in two new copies of this book should count as congruent.

Following common best practice in modern geometry, we solve this problem by defining figures to be *congruent* if there is an isometry taking one onto the other. We define the *degree measure* of an angle to be the least upper bound of an associated set of real numbers. As a result, the assertion that angles are congruent if and only if they have the same degree measure is a theorem that needs to be proven using these two very different definitions. Our Section 5.3 is devoted exclusively to proving that one theorem. Although this book is entirely self contained, a reasonable amount of mathematical background and experience will be called upon to work through it. For those with less mathematical background we suggest considering the text [2] as a more accessible alternative.

We have written this fourth tier book in order to directly anchor K-12 geometry in solid modern mathematics. Our hope is that it will serve as a foundation of fundamental knowledge that will fortify the teaching of secondary and tertiary geometry with deep understanding and inspiration. But this can only happen if mathematicians who are capable of engaging in mathematics at this level will lend their abilities to help elevate the quality of the K-12 mathematics experience.

David M. Clark
Samrat S. Pathania

Acknowledgements

The authors wish to express their gratitude to a number of people whose interest, support and dialog contributed to the successful completion of this book: to Sergei Gelfand for his extended interest in this project, to Peter Renz for his careful feedback on early manuscripts, to our Springer editor Donna Chernyk for her persistence in bringing it to press and to Lorrie Coffey Clark and Laura Wyeth for their forbearance and ever wise counsel.

We would also like to express our gratitude to the many anonymous reviewers whose hard work and thoughtful suggestions have considerably improved the final product.

Contents

Chapter 1
Foundational Principles

This chapter begins our full axiomatic development of the part of plane Euclidean geometry that has traditionally been taught in high school. We will start with a set **P** of objects that we call **points** and a notion of distance between those points. By a **distance function** on **P** we mean a function d from ordered pairs of points into the non-negative real numbers such that

$$d(A, A) = 0 \quad \text{and} \quad d(A, B) = d(B, A)$$

for all points A and B of **P**. We will investigate **planes**

$$\mathcal{P} := \langle \mathbf{P}; d \rangle$$

consisting of a set **P** of points and a distance function d on **P**.

Every physical surface can be modeled by this kind of a plane. Often it is useful to model a plane with different distance functions. For example, the Earth can be modeled as a sphere, where each centimeter on the sphere corresponds to the same number of kilometers on the surface. It can also be modeled by a Mercator projection, where each centimeter alone a single latitude line represents the same number of kilometers. But that number shrinks rapidly as the latitude lines near the poles. In spite of that distortion, Mercator projections have the advantage of describing a flat geometry. Euclidean geometry is the study of flat geometries.

The constructs of *points* and *distance* are thought of as *primitive* or *undefined* constructs because they have no universal meaning. Instead, they are defined in each particular plane. In contrast, a *defined* construct is a construct that is defined using these primitive constructs and previously defined constructs. For example, *between, length, triangle* and *area* will be among the many constructs we will define in general. Defined constructs gain meaning in a particular plane as soon as we define the two primitive constructs in that plane.

Statements of geometry using these constructs are only true or false in a particular plane, where the terms they use have concrete meanings. Any

D. M. Clark, S. Pathania, *A Full Axiomatic Development of High School Geometry*,
https://doi.org/10.1007/978-3-031-23525-2_1

finite set of statements phrased using these constructs can be taken as a set of **axioms** for a theory of geometry. By a **model** of a set of axioms we mean a plane in which each of the axioms is a true statement. A **theorem**, **lemma** or **corollary** of that theory is a further statement that is also true in every model of those axioms.

In the preface of [2] the author explained that he would intentionally omit mention of assumptions about *non-triviality, betweenness* and *intersection* in order to cover the topics of high school geometry in a single undergraduate course. He referred to these assumptions as *Foundational Principles*, and expected that students would fill them in based on their own experiences with physical space as had always been done by students of Euclid. Testimony from instructors using [2] has confirmed that this expectation has been widely realized. These Foundational Principles were first systematically formulated as axioms by David Hilbert [8] in his 1899 revision of Euclid's geometry [4].

In Chapter 1 we will give seven Hilbert style axioms that will replace the Foundational Principles and give two more axioms from [2]. In Chapter 3 we will add one final axiom from [2]. All ten axioms are listed together in Appendix A in the order they will be introduced. The introduction of each axiom will reduce the range of models under consideration, allowing us to say more about the models we most want to study. We will prove theorems from these axioms, and then use those theorems to prove further theorems, gradually building an ever deeper understanding of the properties of these planes.

Chapters 1 through 7 of this book correspond exactly to the first seven chapters of [2], with the order slightly changed and the missing content added. As a result our numbering of lemmas and theorems no longer matches that of [2]. To clarify the correspondence we have added a parenthetic reference to each item from [2]. For example, "[2] C98" after "Theorem 52" means that Theorem 52 is Corollary 98 in [2]. Appendix C: Theorem Index gives a full list of items from [2] with their corresponding item in the present text, indicating where each fact from [2] is fully established.

1.1 Non-triviality and Betweenness

The following initial definitions and axioms illustrate a central rationale for teaching geometry in high school. It gives students an opportunity to work in a context where they can and need to say exactly what they mean and mean exactly what they say. We will model this practice in this text. For example, if we say "A and B are two points", we will have excluded the possibility that they are the same point. But if we say "A is a point and B is a point", we will not have excluded that possibility.

We now give our first defined construct. Point B is **between** points A and C if A, B and C are three points and

$$d(A, C) = d(A, B) + d(B, C).$$

We write $[ABC]$ to mean that B is between A and C. More generally, if $n \geq 3$ is a positive integer and $A_1, A_2, ..., A_n$ are points, we will write

$$[A_1A_2...A_n]$$

as an abbreviation for the conjunction of all statements $[A_iA_jA_k]$ where $i < j < k$. Using the defined construct of *betweenness*, we can immediately give a number of other key defined constructs. If A and B are two points, then the **segment** AB, the **ray** $\overrightarrow{AB}$ and the **line** $\overleftrightarrow{AB}$ are all subsets of $\mathbf{P}$ defined as

$$AB := \{A, B\} \cup \{D \in \mathbf{P} \mid [ADB]\}$$
$$\overrightarrow{AB} := \{A, B\} \cup \{D \in \mathbf{P} \mid [ADB] \text{ or } [ABD]\}$$
$$\overleftrightarrow{AB} := \{A, B\} \cup \{D \in \mathbf{P} \mid [DAB] \text{ or } [ADB] \text{ or } [ABD]\}.$$

Points that lie on the same line are said to be **collinear**.

Using just these definitions see if you can prove that betweenness is symmetric.

Lemma 1 (SYM). *If B is between A and C, then A, B and C are three collinear points and B is between C and A.*

Our first four axioms will serve to eliminate trivial models in which we are likely to have limited interest.

Non-Triviality Axioms

(2PA) Two Point Axiom. *Given two points, there is exactly one line containing them.*

(3PA) Three Point Axiom. *There are at least three points that are not collinear.*

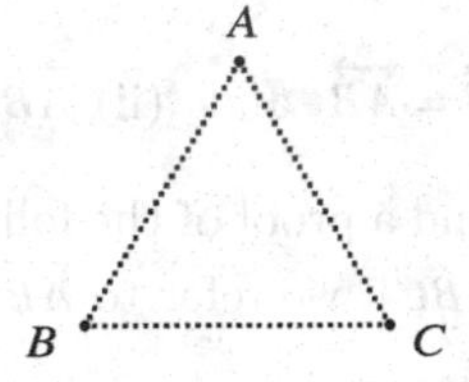

Fig. 1.1 The Three Point Model.

A model of these axioms must contain at least three points. In Figure 1.1 we illustrate a minimal model. Here $\mathbf{P} = \{A, B, C\}$ and the distance between any two points is one. It follows that there are no betweenness relations, and the three 2-element subsets of $\mathbf{P}$ are exactly the segments, the rays and the lines. Not much can be proven from these two axioms alone, as anything that follows from these axioms must be true in this model. While this Three Point Model is not entirely trivial, it is not very interesting, and presses us to add further axioms that will restrict our models to those that better represent physical surfaces. Our first two betweenness axioms capture basic properties we would expect of any notion of betweenness.

Betweenness Axioms

(UNQ) Uniqueness. *Of any three collinear points, exactly one is between the other two.*

(INF) Infinity. *If B and D are two points, then there are points A, C and E such that $[ABCDE]$.*

Note that repeated applications of the INF axiom will produce an unlimited number of points on the line $\overleftrightarrow{BD}$. The axiom UNQ is vacuously true in the Three Point Model since this model does not contain three collinear points. But the INF axiom fails completely in that model, eliminating it from further consideration. However, all four axioms we have given hold in every real Cartesian space $\mathbb{R}^n$, where n is a positive integer greater than one. As an application of UNQ and INF, you can establish an important property of the distance function.

Lemma 2. *If $A, B \in \mathbf{P}$, then $d(A, B) = 0$ if and only if $A = B$.*

You can prove the next lemma by applying the definition of *ray* together with SYM and UNQ. It will follow that the statement of this lemma is true in every model of the axioms we have given. We will use "$\overleftarrow{AB}$" as an alternative name for $\overrightarrow{BA}$.

Lemma 3. *If A and B are two points, then*

$$\text{(i) } \overleftarrow{AB} \cup \overrightarrow{AB} = \overleftrightarrow{AB} \quad \text{and} \quad \text{(ii) } \overleftarrow{AB} \cap \overrightarrow{AB} = AB.$$

You might like to try to find a proof of the following similar looking lemma before reading further. If $[ABC]$, we refer to $\overleftarrow{AB}$ and $\overrightarrow{BC}$ as **opposite rays**.

Lemma X. *If $[ABC]$ for three points A, B and C on line ℓ, then*

$$\text{(i) } \overleftarrow{AB} \cup \overrightarrow{BC} = \ell \quad \textit{and} \quad \text{(ii) } \overleftarrow{AB} \cap \overrightarrow{BC} = \{B\}.$$

If you cannot prove it now, you can come back to it when you have more axioms.

Since we are interested in *plane geometry*, we will now add an axiom that holds in the real coordinate plane $\mathbb{R}^2$ but fails in $\mathbb{R}^n$ for all n greater than two. Let ℓ be a line and let A and C be points not on ℓ. We say that A and C are on the **same side** of ℓ if no point of ℓ is between them. (Figure 1.2.)

(PSA) Plane Separation Axiom. *Let ℓ be a line and let A, B and C be three points not lying on ℓ.*

(i) *If A and B are on the same side of ℓ and B and C are on the same side of ℓ, then A and C are on the same side of ℓ.*

(ii) *If A and B are not on the same side of ℓ and B and C are not on the same side of ℓ, then A and C are on the same side of ℓ.*

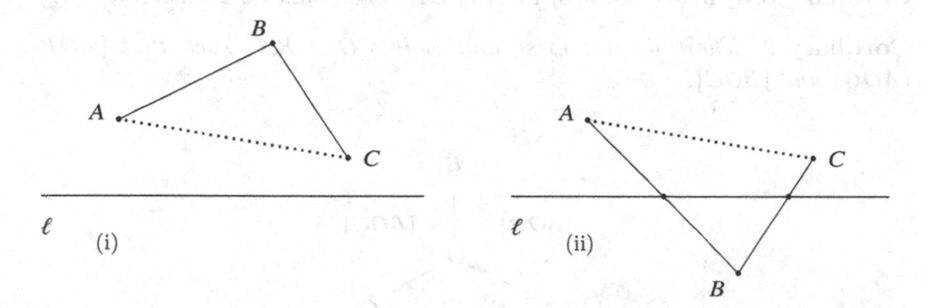

Fig. 1.2 Sides of a line ℓ and the PSA.

Theorem 4. *If ℓ is a line, then being on the same side of ℓ is an equivalence relation on the set of points not on ℓ that has exactly two equivalence classes.*

By a **side** of line ℓ we mean one of the two equivalence classes of this equivalence relation. Thus, if E and F are not on ℓ, then they are either on the same side of ℓ or they are on opposites sides of ℓ. As a result ℓ partitions the plane into three parts, itself and its two sides.

If A, B and C are three non-collinear points, then the **triangle** $\triangle ABC$ is defined as

$$\triangle ABC := AB \cup BC \cup AC.$$

The points A, B and C are called the **vertices** of $\triangle ABC$ and the segments AB, BC and AC are called the **sides** of $\triangle ABC$.

Pasch's Theorem 5. *Let ℓ be a line that does not contain any of the vertices of $\triangle ABC$. If ℓ intersects one of the sides of $\triangle ABC$, then it intersects one and only one of its other two sides.*

Lemma 6. *If point A is on a line ℓ and point B is not on ℓ, then every point of the ray $\overrightarrow{AB}$ other than A is on the B-side of ℓ.*

The following theorem tell us more about how points are arranged on a single line.

Theorem 7. *Let A, B, C and D be points.*

(i) **(4Pt1).** *If $[ABC]$ and $[ACD]$, then $[ABCD]$.*
(ii) **(4Pt2).** *If $[ABC]$ and $[BCD]$, then $[ABCD]$.*

(Both 4Pt1 and 4Pt2 concern only points on a single line. To prove them, look at the sides of some second line that intersects this line.)
We have given special names to the two parts of Theorem 7 because they will be used extensively in all that follows.

Corollary 8. *If B is a point of the ray $\overrightarrow{OA}$ other than O, then $\overrightarrow{OB} = \overrightarrow{OA}$.*

Corollary 9. *There do not exist four points O, A, B, C such that $[AOB]$, $[AOC]$ and $[BOC]$.*

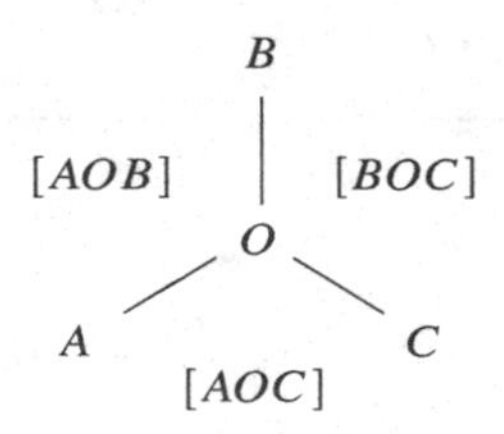

Fig. 1.3 Forbidden quadruple.

At last we have what we need to prove Lemma X. Part (i) will use both parts of Theorem 7 and part (ii) will use Corollary 9.

Lemma 10. *If $[ABC]$ for three points A, B and C on line ℓ, then*

$$\text{(i) } \overleftarrow{AB} \cap \overrightarrow{BC} = \{B\} \quad \textit{and} \quad \text{(ii) } \overleftarrow{AB} \cup \overrightarrow{BC} = \ell.$$

If A, O and B are three non-collinear points, then the **angle** $\angle AOB$ is defined as the union

$$\angle AOB := \overrightarrow{OA} \cup \overrightarrow{OB}$$

of the rays $\overrightarrow{OA}$ and $\overrightarrow{OB}$. If the two rays are clear from the context, we will sometimes refer to this as $\angle O$. The **interior** of the angle $\angle AOB$ is the set of all points on the A-side of $\overrightarrow{OB}$ and on the B-side of $\overrightarrow{OA}$.

We next have another theorem that will be used sufficiently often to give it a special name.

Crossbar Theorem 11. *If point D is in the interior of $\angle AOB$, then ray $\overrightarrow{OD}$ intersects AB at a point between A and B.*

(Consider Figure 1.4 where we have added points E and F as shown. Use Pasch's Theorem to conclude that $\overleftrightarrow{OD}$ intersects one of the other two sides of $\triangle AEB$ at some point X. Then use the information you have about sides of lines to show that X is on $\overrightarrow{OD}$ and is between A and B.)

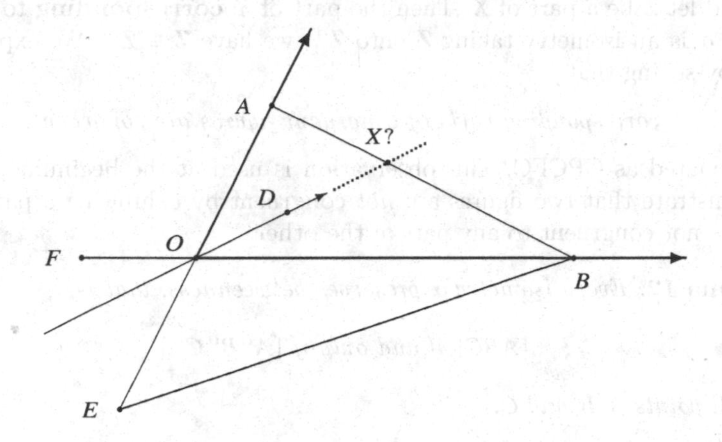

Fig. 1.4 Crossbar Theorem.

1.2 Congruence and Length Measure

An **isometry** of a plane $\mathcal{P}$ is a bijection $\alpha : \mathbf{P} \to \mathbf{P}$ that takes each point A to the point A^α in such a way that, for all $A, B \in \mathbf{P}$,

$$d(A^\alpha, B^\alpha) = d(A, B).$$

Note that the composition of two isometries is an isometry, the identity bijection, $A \mapsto A$, is an isometry and the inverse of an isometry is an isometry. Consequently the set of all isometries of $\mathcal{P}$ forms the group of its distance preserving bijections under composition.

A subset of $\mathbf{P}$ is called a **figure** or **region**. Figures $\mathbf{X}$ and $\mathbf{Y}$ are **congruent**, written

$$\mathbf{X} \cong \mathbf{Y},$$

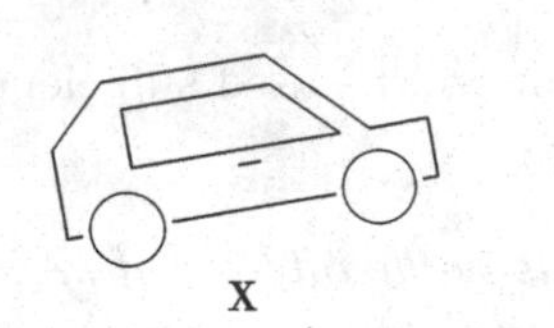

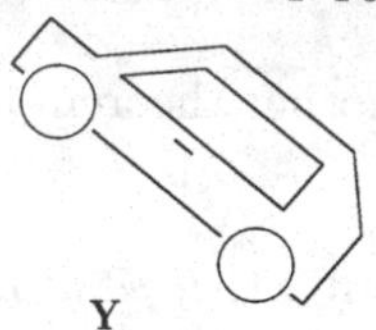

Fig. 1.5 Congruent figures **X** and **Y**. [2]

if there is an isometry that takes **X** onto **Y**. (See Figure 1.5.) If **X** and **Z** are figures, then **Z** is a **part** of **X** if $\mathbf{Z} \subseteq \mathbf{X}$. Let α be an isometry taking **X** onto **Y**, and let **Z** be a part of **X**. Then the part of **Y** **corresponding** to **Z** is $\mathbf{Z}^\alpha$. Since α is an isometry taking **Z** onto $\mathbf{Z}^\alpha$, we have $\mathbf{Z} \cong \mathbf{Z}^\alpha$. We express this fact by saying that

corresponding parts of congruent figures are congruent,

abbreviated as CPCFC. This observation is used at the beginning of [2] to demonstrate that two figures are *not* congruent by exhibiting a part of one that is not congruent to any part of the other.

Lemma 12. *Every isometry α preserves betweenness, that is,*

$$[ABC] \text{ if and only if } [A^\alpha B^\alpha C^\alpha]$$

for all points A, B and C.

The proof of Lemma 12 points to a general principle. Ultimately every defined construct, like betweenness, can be defined in terms of points and distance alone. Since an isometry is a bijection on points and preserves distance, we should expect it to *preserve* every defined construct in an appropriate sense for that construct.

It turns out that our primary direct use of the distance function d is to measure the lengths of segments. Our intent is to define the length of a segment as the distance between its endpoints. Doing this requires that we first verify that its endpoints are uniquely determined, that is, $AB = CD$ implies $\{A, B\} = \{C, D\}$.

Lemma 13. *If A and B are two points, then they are the only points of segment AB that are not between any other two points of AB.*

This lemma allows us to unambiguously define *the* **endpoints** of a segment to be the two points of that segment not between two other points of that segment. Also the *distance between the endpoints of AB* is unambiguously defined since $d(A, B) = d(B, A)$. We can now define the **length** of AB as the distance between its endpoints:

$$\mathcal{L}(AB) := d(A, B).$$

By Lemma 2 the **length measure function** $\mathcal{L}$ is a function from the set of all segments into the *positive* real numbers.

The definitions of *betweenness* and *isometry* can also be expressed in terms of length measure. Point B is **between** points A and C if and only if A, B and C are three points and

$$\mathcal{L}(AC) = \mathcal{L}(AB) + \mathcal{L}(BC).$$

A bijection α of **P** is an **isometry** if and only if, for every segment AB,

$$\mathcal{L}(A^\alpha B^\alpha) = \mathcal{L}(AB).$$

Our next axiom gives properties of length measure that relate it to the number system, to congruence and to the sides of a triangle.

(LMA) Length Measure Axiom. *The length measure function* $\mathcal{L}$ *has the following properties.*

(i) *There are two points* O *and* I *such that* $\mathcal{L}(OI) = 1$.
(ii) *Segments are congruent if and only if they have the same length.*
(iii) $\mathcal{L}(AC) < \mathcal{L}(AB) + \mathcal{L}(BC)$ *for any three non-collinear points* A, B *and* C (the **Triangle Inequality**).

This is a modification of Axiom 1 of [2] where the Triangle Inequality has replaced Axiom 1(iii), now the definition of betweenness. This inequality is not mentioned in [2], but follows easily from Corollary 98 of [2] by looking at the altitude from base AC to vertex B.

The often used Ruler Postulate of Birkhoff [1] implies that, for every positive real number, there is a segment having that number as its length. (See page 100.) Following [2] we do not use this postulate as it is not needed in our development. Instead we use LMA(i), a very small fragment of the Ruler Postulate. We will see in Chapter 8 that this choice results in a much richer assortment of models of our geometry, including the coordinate planes of the constructible numbers and of the algebraic numbers. We refer to the set of all positive real numbers that are lengths of segments as $\mathbb{F}^+$. Thus

$$\mathbb{F}^+ := \{\mathcal{L}(AB) \mid A, B \in \mathbf{P} \text{ with } A \neq B\} \subseteq \mathbb{R}^+,$$

where $\mathbb{R}^+$ denotes the set of all positive real numbers. We note here that readers do have the option to include the Ruler Postulate if they like. This will mean that the number set $\mathbb{F}^+$ is necessarily all of $\mathbb{R}^+$. As a result all parts of our development that concern which numbers are in $\mathbb{F}^+$ could be ignored. This would include most of Chapter 8 where we exhibit a wide range of models of our axioms that correspond to different subsets $\mathbb{F}^+$ of $\mathbb{R}^+$.

1.3 Circles

In this section we will add two axioms about intersections. If O is a point and $r \in \mathbb{F}^+$, then the set

$$\mathbf{C}_r^O := \{B \in \mathbf{P} \mid d(O, B) = r\}$$

is called a **circle**. The point O is the **center** of **C** and r is **the radius** of **C**. We require the radius to be in $\mathbb{F}^+$ because, if it were not, the circle would be the empty set. Our next axiom below will imply that this provision guarantees that the circle is not the empty set. The name "$\mathbf{C}_r^O$" for a circle indicates that O is its center and that r is its radius. Rather than naming the radius, we will often specify a circle as "$\mathbf{C}_B^O$" to indicate that B is a point on it and that its radius is therefore $r = d(O, B)$.

For each point B on circle $\mathbf{C}_r^O$, we say that the segment OB is **a radius** of $\mathbf{C}_r^O$. If OA and OB are radii and O is between A and B, then AB is **a diameter** of $\mathbf{C}_r^O$. By the definition of *between* all diameters have the same length, namely $2r$. We call the number $2r$ **the diameter of circle** $\mathbf{C}_r^O$. The **inside** and **outside** of circle $\mathbf{C}_r^O$ are defined as

$$\mathrm{In}(\mathbf{C}_r^O) = \{X \in \mathbf{P} \mid d(O, X) < r\}$$
$$\mathrm{Out}(\mathbf{C}_r^O) = \{Y \in \mathbf{P} \mid d(O, Y) > r\}.$$

Lemma 14. *Let A, B and O be three points. Then*

(i) *B lies outside circle $\mathbf{C}_A^O$ if $[OAB]$ and*
(ii) *B lies inside circle $\mathbf{C}_A^O$ if $[OBA]$.*

(RCA) Ray Circle Axiom. *A circle intersects every ray emanating from a point inside the circle.*

Theorem 15. *If a point of line ℓ is inside a circle, then it is between two points of ℓ that are on the circle.*

Corollary 16. *Every circle has a diameter.*

Because no precise definition of *congruent* is given in [2], the only figures that are ever provably congruent are triangles and parts of triangles. The transformational definition of *congruent* using isometries opens the possibility of proving other figures congruent.

Theorem 17. *Two circles are congruent if and only if they have the same radius.*

The RCA tells us something about how rays, and therefore also lines, will intersect circles. We will need a second circle axiom that tells us how circles intersect each other.

(CCA) Circle Circle Axiom. *If a circle contains a point outside and a point inside another circle, then the two circles intersect in exactly two points which lie on opposite sides of the line containing their centers.*

1.4 Triangle Congruence

Consider two triangles $\triangle ABC$ and $\triangle XYZ$. We will write

$$\triangle ABC \cong \triangle XYZ$$

to mean that these triangles are congruent under an isometry α such that $A^\alpha = X$, $B^\alpha = Y$ and $C^\alpha = Z$. Conversely, if α is an isometry taking A, B, C to X, Y, Z, respectively, then $(\triangle ABC)^\alpha = \triangle XYZ$ by Lemma 12, and therefore $\triangle ABC \cong \triangle XYZ$.

A triangle has six parts: three angles and three sides. If two triangles are congruent, then CPCFC tells us that their respective parts must be congruent under the associated isometry. It is natural to ask if the converse is true. Assume that we have two triangles $\triangle ABC$ and $\triangle XYZ$ for which the three corresponding sides and three corresponding angles are pairwise congruent:

$$AB \cong XY, \; BC \cong YZ, \; CA \cong ZX$$
$$\angle ABC \cong \angle XYZ, \; \angle BCA \cong \angle YZX, \; \angle CAB \cong \angle ZXY.$$

Does it follow that the two triangles are congruent?

To answer this question affirmatively would require finding an isometry of the *entire plane* that takes $\triangle ABC$ onto $\triangle XYZ$. At this point we do not have the means to do this. Instead we will take as an axiom the assertion that the first three of these conditions imply congruence. Using this axiom we will eventually demonstrate that a number of different other subsets of these six congruences also imply that the triangles are congruent. In this section we will prove the most important instance of this, the SAS axiom of [2].

(SSS) Side-Side-Side. *If $\triangle ABC$ and $\triangle XYZ$ are triangles with $AB \cong XY$, $AC \cong XZ$ and $BC \cong YZ$, then $\triangle ABC \cong \triangle XYZ$.*

Lemma 18. *If A and B are two points, then A is the only point of $\overrightarrow{AB}$ that is not between two other points of $\overrightarrow{AB}$.*

Lemma 19. *If φ is an isometry taking $\angle ABC$ onto $\angle XYZ$, then $B^\varphi = Y$ and $\{(\overrightarrow{BA})^\varphi, (\overrightarrow{BC})^\varphi\} = \{\overrightarrow{YX}, \overrightarrow{YZ}\}$.*

Lemma 20 ([2] T29). *Given a segment AB and a ray $\overrightarrow{CD}$, there is a unique point X on $\overrightarrow{CD}$ such that $AB \cong CX$.*

Lemma 21. *Let $\triangle ABC$ and $\triangle XYZ$ be triangles such that $AB \cong XY$, that $BC \cong YZ$ and that there is an isometry φ taking $\angle ABC$ onto $\angle XYZ$. Then $AC \cong XZ$.*

(Figures 1.6 and 1.7 suggest two ways that φ might take $\angle ABC$ onto $\angle XYZ$.)

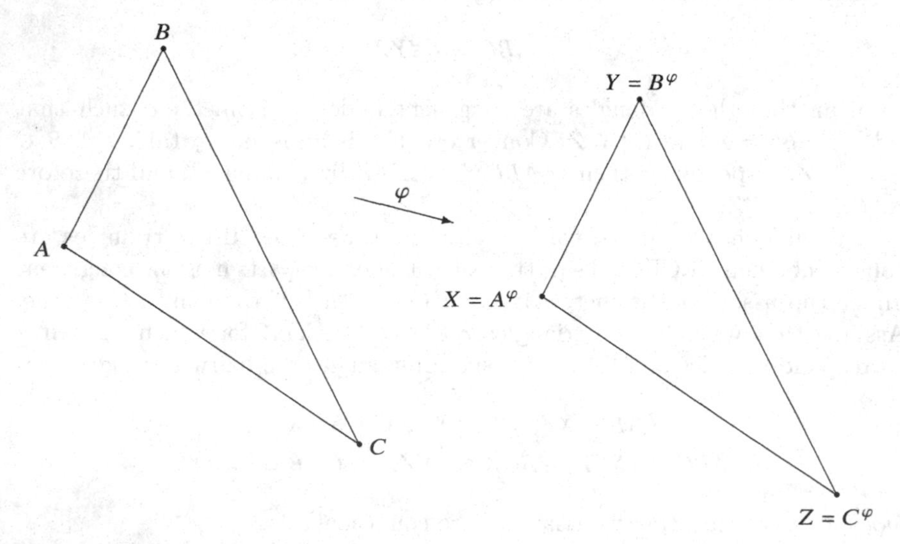

Fig. 1.6 $\mathcal{L}(AC) = \mathcal{L}(XZ)$, Case 1.

Theorem 22 ([2] (**Axiom SAS**). *If $\triangle ABC$ and $\triangle XYZ$ are triangles with $AB \cong XY$, $\angle ABC \cong \angle XYZ$ and $BC \cong YZ$, then $\triangle ABC \cong \triangle XYZ$.*

1.5 Upgrading Euclidean Constructions

Plane geometry is a study of the various figures that lie within a plane. Figures are discussed using carefully defined terms to describe their properties. Here we give a number of important examples.

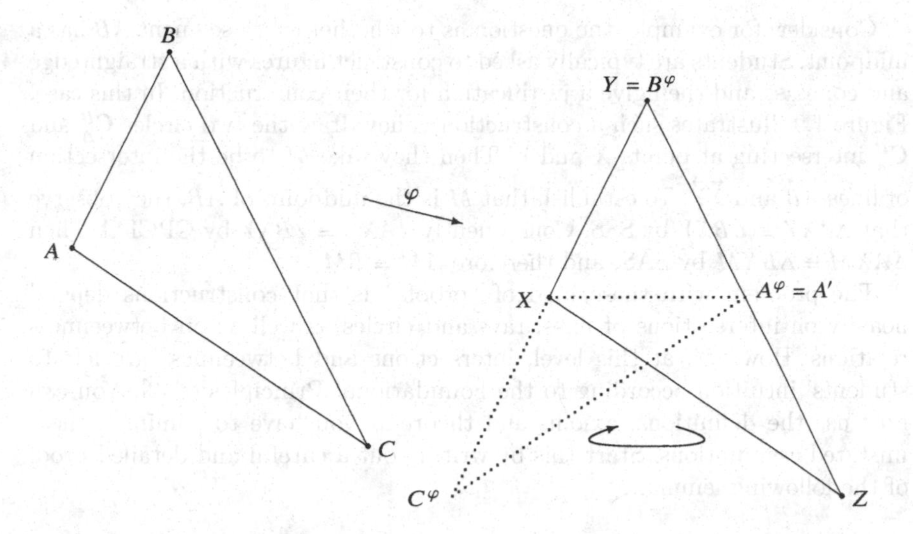

Fig. 1.7 $\mathcal{L}(AC) = \mathcal{L}(XZ)$, Case 2.

Let A, B and C be collinear points with B between A and C, and let D be a point not on the line containing A, B and C. Then the angles $\angle ABD$ and $\angle CBD$ are called **supplementary angles**. A **right angle** is an angle that is congruent to one of its supplements. We say that two intersecting lines l and m are **perpendicular** if they form a right angle at their intersection. An **equilateral triangle** is a triangle whose three sides are pairwise congruent. A **trisquare** is a triangle that has three right angles. (Figure 1.8.) Given a segment AB, we say that a point M between A and B is a **midpoint** of AB if $AM \cong MB$. Ray $\overrightarrow{OM}$ **bisects** $\angle AOB$ if M in in the interior of $\angle AOB$ and $\angle AOM \cong \angle MOB$.

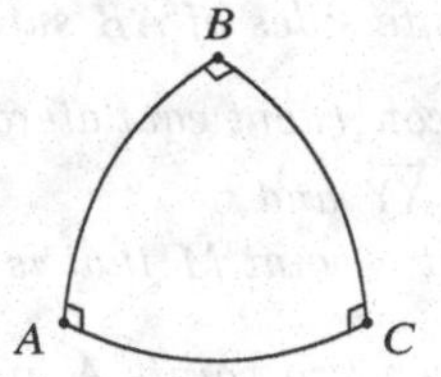

Fig. 1.8 A trisquare.

Defined terms like *equilateral triangle*, *trisquare*, *midpoint* and *bisector* raise an obvious question:

Do these objects necessarily exist in the plane $\mathcal{P}$*?*

Looking at how this question is answered in [2] offers a nice illustration of the benefits of our work in this chapter.

Consider, for example, the question as to whether every segment AB has a midpoint. Students are typically asked to construct figures with a straightedge and compass and then give a justification for their construction. In this case, Figure 1.9 illustrates such a construction. They draw the two circles $\mathbf{C}_A^B$ and $\mathbf{C}_A^B$ intersecting at points X and Y. Then they take M to be the intersection of lines $\overleftrightarrow{AB}$ and $\overleftrightarrow{XY}$. To establish that M is the midpoint of AB, they observe that $\triangle AXY \cong \triangle BXY$ by SSS. Consequently $\angle AXY \cong \angle BXY$ by CPCFC. Then $\triangle AXM \cong \triangle BXM$ by SAS, and therefore $AM \cong BM$.

The problem with these kind of "proofs" is that constructions depend heavily on intersections of lines, rays and circles as well as on betweenness relations. However, at this level, intersections and betweenness are left to students' intuition according to the Foundational Principles of [2]. You can now use the definitions, axioms and theorems you have to eliminate these unstated assumptions. Start this by writing out a careful and detailed proof of the following lemma.

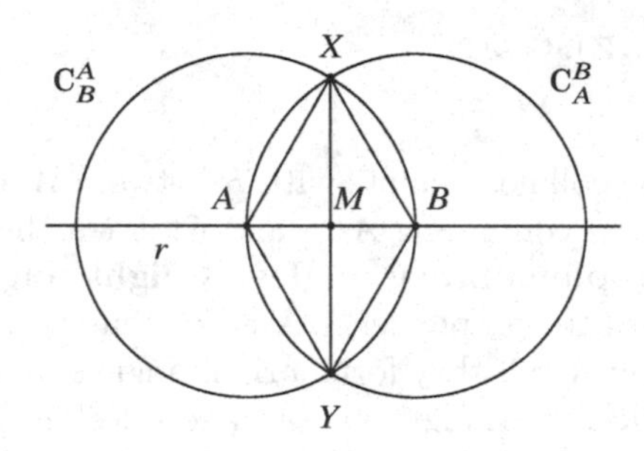

Fig. 1.9 Midpoint of segment AB.

Lemma 23. *If A and B are two points, then circles $\mathbf{C}_A^B$ and $\mathbf{C}_B^A$ intersect in two points X and Y on opposite sides of $\overleftrightarrow{AB}$ such that*

(i) *$\triangle ABX$ and $\triangle ABY$ are congruent equilateral triangles,*
(ii) *$\overleftrightarrow{AB}$ is perpendicular to $\overleftrightarrow{XY}$ and*
(iii) *AB and XY intersect at a point M that is a midpoint of both.*

Corollary 24 ([2] T30). *Given two points A and B, there is a third point C such that $\triangle ABC$ is an equilateral triangle.*

Theorem 25 ([2] T35). *Every segment has a unique midpoint.*

Theorem 26 ([2] T31). *Given an angle $\angle ABC$, a ray $\overrightarrow{EF}$, and a point Y not on line $\overleftrightarrow{EF}$, there is a point X on the Y-side of line $\overleftrightarrow{EF}$ so that $\angle ABC \cong \angle XEF$.*

(See Figure 1.10).

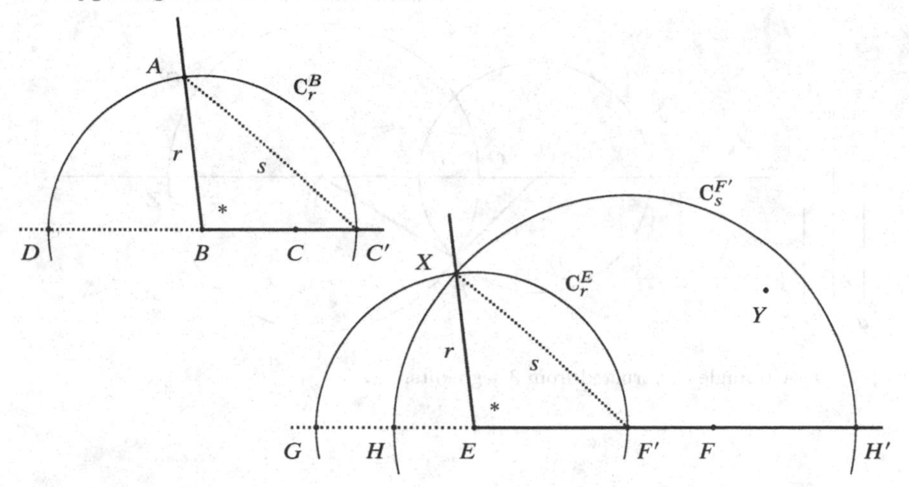

Fig. 1.10 Copying $\angle ABC$.

You are now in a position to establish the existence of perpendiculars and angle bisectors defined at the beginning of this section.

Theorem 27 ([2] T34). *There is a line perpendicular to any given line through a given point on the line.*

Theorem 28 ([2] T33). *There is a line perpendicular to any given line through a given point not on the line.*

Theorem 29 ([2] T32). *Every angle has a unique bisector.*

Note that the foundation we have developed here allows us to add "unique" to Theorems 29, 32 and 35 of [2], justifying references to *the* midpoint of a segment and *the* bisector of an angle.

Problem 20 in [2] asks students to construct a triangle with sides congruent to three given segments. Of course, those three segments were chosen so that such a triangle could be constructed. This problem leads us to ask: For which triples of segments does there exist a triangle with sides congruent to those three segments? We now have the facts we need to answer this question.

Theorem 30. *Let a, b and c be the lengths of three segments. Then there is a triangle with sides of length a, b and c if and only if each of these numbers is less than the sum of the other two.*

(See Figure 1.11.)

It still remains to answer the question as to whether or not the plane contains a trisquare.

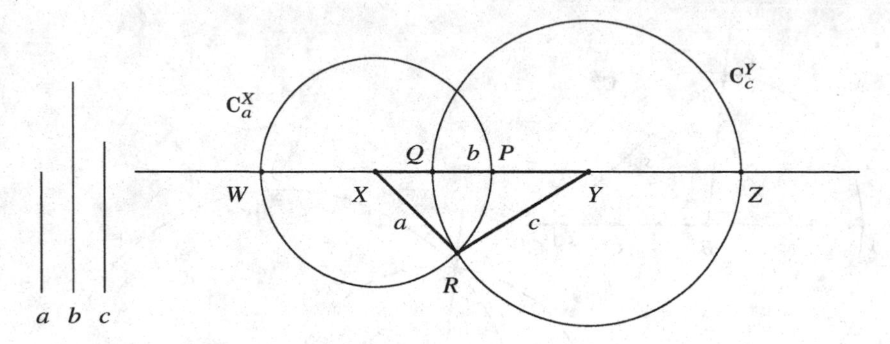

Fig. 1.11 A triangle constructed from 3 segments.

Chapter 2
Neutral Geometry

Two lines are **parallel** if they have no point in common. It turns out that many of the properties of a plane $\mathcal{P}$ that we would like to understand depend on the answer to a simple question:

Given a line ℓ and a point P not on ℓ, how many lines parallel to ℓ pass through P?

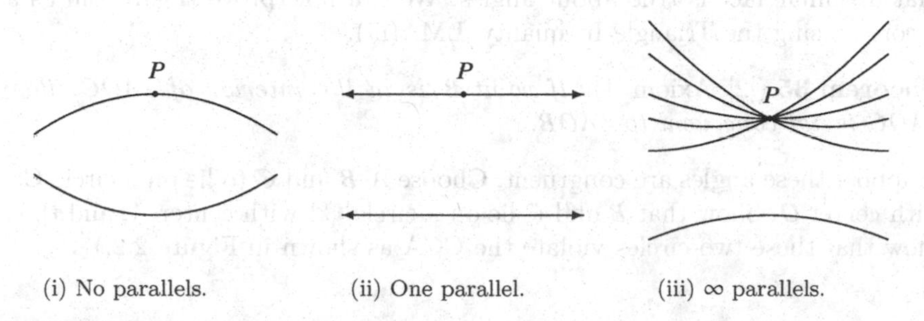

Fig. 2.1 Existence of parallels.

The answer will depend on what plane we have. It is conceivable that all pairs of lines would bend toward each other and eventually intersect, as shown in Figure 2.1(i). Or there might be a plane in which all lines bend away from each other, as shown in Figure 2.1(iii), resulting in infinitely many lines through P parallel to ℓ. We will not study either of these alternative planes in this book, but only mention that there is ample literature written about both.

In the next chapter we will add one final axiom that says that the answer is always "exactly one line", as shown in Figure 2.1(ii). In the meantime, we will see in this chapter that there are many important facts about geometry that can be proven without any assumption about the number of parallels through a point P. As you prove these theorems, pay close attention to places where you need to know which points are between which other points, which

D. M. Clark, S. Pathania, *A Full Axiomatic Development of High School Geometry*,
https://doi.org/10.1007/978-3-031-23525-2_2

lines and circles intersect and which side of a line you are on. The facts proven in Chapter 1 will help you through this.

We begin by looking at a special kind of triangle. A triangle is **isosceles** if two of its sides are congruent segments. The angles opposite the congruent sides of an isosceles triangle are called the **base angles**.

Theorem 31 ([2] T36). *The base angles of an isosceles triangle are congruent angles.*

Theorem 32 ([2] T37). *Congruent angles have congruent supplements.*

Assume that X is a point between A and C, that X is also between B and D, and that these points are not all collinear. Then the angles $\angle AXB$ and $\angle CXD$ are called **vertical angles**.

Theorem 33 ([2] C38). *Vertical angles are congruent.*

Weak Right Angle Theorem 34 ([2] T39). *An angle that is congruent to a right angle is also a right angle.*

If $[ABC]$, then $\mathcal{L}(AC) = \mathcal{L}(AB) + \mathcal{L}(BC)$. It follows that $\mathcal{L}(AC) > \mathcal{L}(AB)$ and consequently AC is not congruent to AB by LMA(ii). Axiom 4 of [2] says that a similar fact is true about angles. We can now prove this axiom as a theorem using the Triangle Inequality, LMA(iii).

Theorem 35 ([2] Axiom 4). *If point B is in the interior of $\angle AOC$, then $\angle AOC$ is not congruent to $\angle AOB$.*

(Suppose these angles are congruent. Choose A, B and C to lie on a circle $\mathbf{C}^O$ with center O. Show that B and C lie on a circle $\mathbf{C}^A$ with center A, and then show that these two circles violate the CCA as shown in Figure 2.2.)

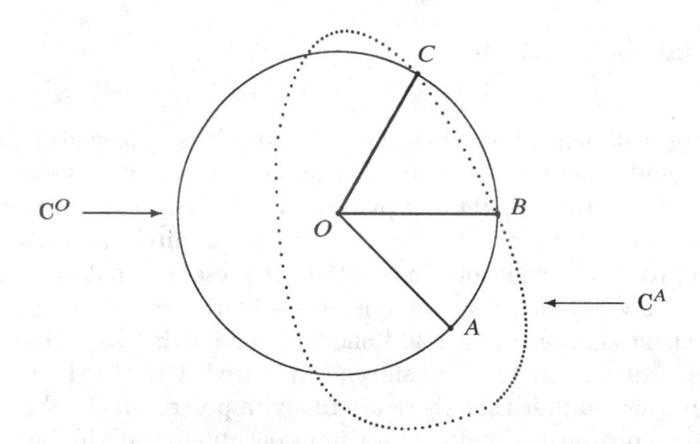

Fig. 2.2 Proof of Theorem 35.

We can at last settle the matter of the trisquares.

Theorem 36 ([2] T47). *A triangle has at most one right angle. In particular, there are no trisquares.*

Figure 2.1(i) suggests correctly that surfaces with uniform positive curvature, like a sphere, can contain trisquares.

Assume that a line ℓ intersects two lines m and n at two different points. The line ℓ is called a **transversal** to m and n. The angles in Figure 2.3 marked $\diamond$ and $*$ are called **corresponding angles** and the angles marked $\circ$ and $*$ are called **alternate interior angles**.

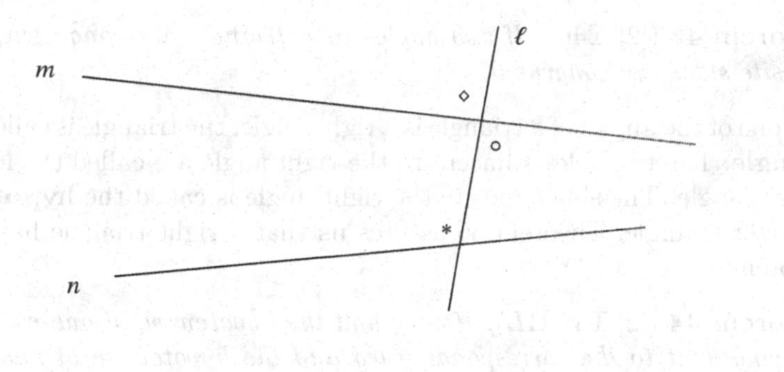

Fig. 2.3 Alternate interior angles and corresponding angles. [2]

Theorem 37 ([2] T40). *If two lines have a transversal which forms alternate interior angles that are congruent, then the two lines are parallel.*

Corollary 38 ([2] C41). *If two lines have a transversal which forms corresponding angles that are congruent, then the two lines are parallel.*

Weak Parallel Line Corollary 39 ([2] C42). *Given a line ℓ and a point P not on ℓ, there is a line m such that P is on m and m is parallel to ℓ.*

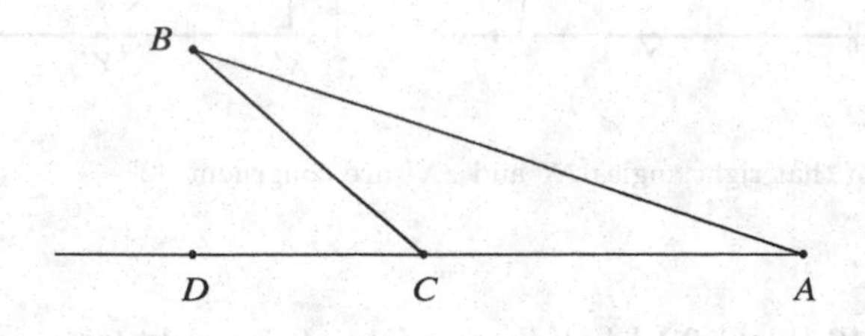

Fig. 2.4 An angle exterior to $\triangle ABC$. [2]

Given $\triangle ABC$ and point D such that C is between A and D, the angle $\angle BCD$ is called an **exterior angle** of $\triangle ABC$. Angles $\angle A$ and $\angle B$ are called the **opposite interior angles** to $\angle BCD$. (Figure 2.4.)

Weak Exterior Angle Theorem 40 ([2] T43). *An exterior angle of a triangle is not congruent to either opposite interior angle.*

Theorem 41 ([2] T44 **ASA**). *If two angles and the side between them in one triangle are congruent to the corresponding two angles and side between them in another triangle, then the triangles themselves are congruent.*

Theorem 42 ([2] T45 **AAS**). *If two angles and a side not between them in one triangle are congruent to the corresponding two angles and side not between them in another triangle, then the triangles themselves are congruent.*

Theorem 43 ([2] T46). *If two angles of a triangle are congruent, then the opposite sides are congruent.*

If one of the angles of a triangle is a right angle, the triangle is called a **right triangle**. The two sides adjacent to the right angle are called the **legs** of the right triangle. The side opposite the right angle is called the **hypotenuse** of the right triangle. Theorem 36 assures us that a right triangle has only one hypotenuse.

Theorem 44 ([2] T48 **HL**). *If a leg and the hypotenuse of one right triangle are congruent to the corresponding leg and the hypotenuse of another right triangle, then the two right triangles are congruent.*

Strong Right Angle Theorem 45 ([2] T49). *All right angles are congruent. Thus a right angle is congruent to another angle if and only if the other angle is also a right angle* (Figure 2.5).

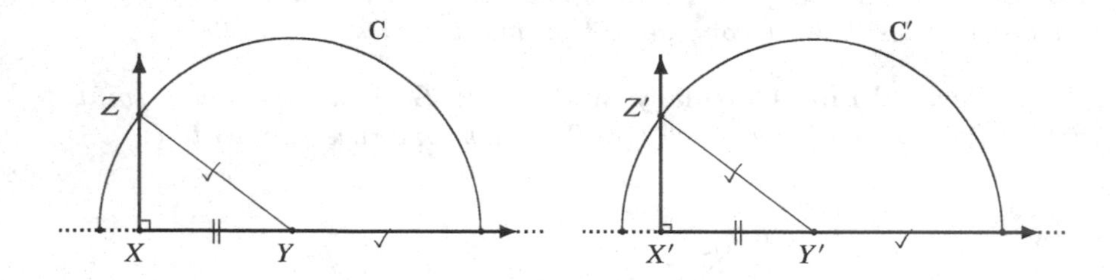

Fig. 2.5 Proof that right angles $\angle X$ and $\angle X'$ are congruent. [2]

Corollary 46 ([2] C50 **LL**). *If the two legs of one right triangle are congruent to the corresponding two legs of another right triangle, then the two right triangles are congruent.*

Corollary 47 ([2] C51 **HA**). *If the hypotenuse and a non-right angle of one right triangle are congruent to the hypotenuse and a non-right angle of another right triangle, then the two right triangles are congruent.*

Corollary 48 ([2] C52 **LA**). *If a leg and non-right angle of one right triangle are congruent to the corresponding leg and non-right angle of another right triangle, then the two right triangles are congruent.*

A **tangent** to the circle **C** is a line that contains exactly one point of **C**.

Theorem 49 ([2] T53). *Let ℓ be a line that contains a point T of the circle* **C** *with center O. Then ℓ is tangent to* **C** *if and only if the radius OT is perpendicular to ℓ.*

Theorem 50 ([2] P55). *If P is a point outside circle* **C** *with center O, then there are points T_1 and T_2 of* **C** *on opposite sides of $\overleftrightarrow{OP}$ such that $\overleftrightarrow{PT_1}$ and $\overleftrightarrow{PT_2}$ are both tangent to* **C**. (Figure 2.6.)

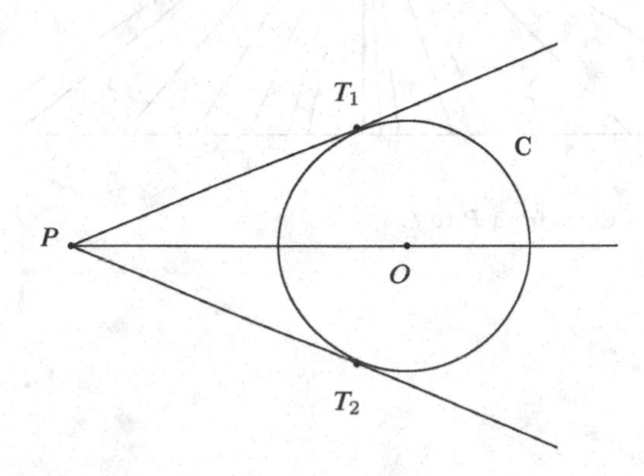

Fig. 2.6 Tangents to a circle.

Theorem 51 ([2] T54). *Let P be a point outside of the circle* **C** *with center O and let T and T' be two points on* **C** *such that $\overleftrightarrow{PT}$ and $\overleftrightarrow{PT'}$ are both tangent to* **C**. *Then $PT \cong PT'$.*

In the sixth century BCE the Greek mathematician and philosopher Pythagoras of Samos recorded a theorem that has remained central to plane Euclidean geometry to this day. It can be a valuable exercise for high school students to investigate the many known proofs of Pythagoras' theorem. This theorem is unusual in the sense that new insights into geometry often lead to new proofs of this theorem.

We will present three different versions of this theorem with five different proofs. Each will reflect what can be done with the tools available at that point in our development. We start here with a very weak version that will be used to prove the full version at the end of Chapter 3. In Chapter 3 we will also give a historically significant version that dates back to early Greece. Chapters 4 and Chapter 5 will each end with a proof of the full version that utilizes what we will have learned in that chapter.

1st Pythagorean Theorem 52 ([2] C98). *The hypotenuse of a right triangle is longer than either of its legs.*

We can rephrase this theorem by saying that, given a point P and a line ℓ not containing P, the perpendicular segment from P to ℓ is shorter than any other segment connecting P to a point on ℓ (Figure 2.7).

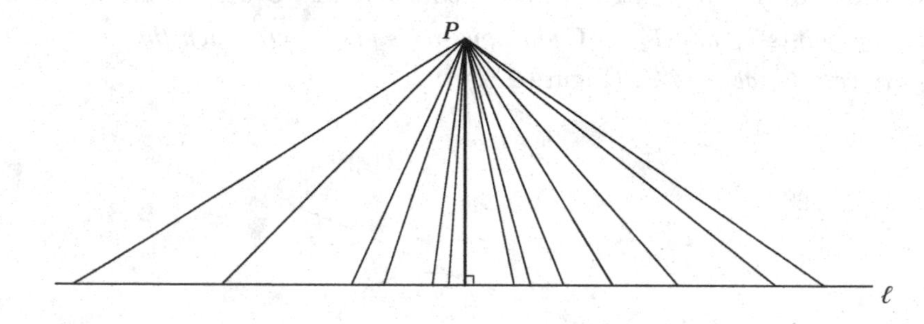

Fig. 2.7 Shortest distance from P to ℓ.

Chapter 3
Similar Figures

Fix any point O. If x is a positive number, we would like to define the *dilation* with *center* O and *scaling factor* x to be the function taking O to itself and, for each point $P \neq O$, taking P to the point xP on ray $\overrightarrow{OP}$ such that $d(O, xP) = xd(O, P)$. This dilation is illustrated in Figure 3.1.

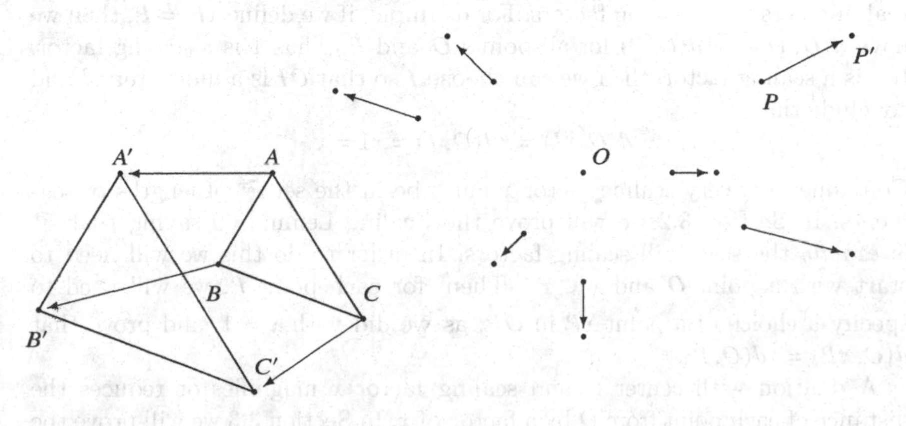

Fig. 3.1 Dilation from O with scaling factor $k = \frac{3}{2}$. [2]

The problem with this definition is that it requires that we first show that *there is* such a point xP on $\overrightarrow{OP}$. To resolve this problem we will start in a different place. We say that a positive real number x is a **scaling factor** if, for any two points O and P, there is a (necessarily unique) point xP on ray $\overrightarrow{OP}$ such that

$$d(O, xP) = xd(O, P).$$

In this case we refer to the function $O \mapsto O$ and $P \mapsto xP$ as the **dilation** with **center** O and **scaling factor** x. We will say two figures $\mathbf{X}$ and $\mathbf{Y}$ are

D. M. Clark, S. Pathania, *A Full Axiomatic Development of High School Geometry*,
https://doi.org/10.1007/978-3-031-23525-2_3

similar if one is congruent to the image of the other under some dilation. This is expressed as

$$\mathbf{X} \sim \mathbf{Y},$$

where we use the symbol "$\sim$" for "is similar to".

Fig. 3.2 Similar figures.

In order to make use of this concept, we need to determine which positive real numbers x are scaling factors. For example, if we define $1P := P$, then we have $d(O, 1P) = 1d(O, P)$ for all points O and P. Thus 1 is a scaling factor. If x is a scaling factor, then we can choose I so that OI is a unit interval and conclude that

$$d(O, xI) = xd(O, I) = x1 = x.$$

Consequently every scaling factor x must be in the set $\mathbb{F}^+$ of lengths of segments. In Section 3.2 we will prove the Scaling Lemma 69 saying that $\mathbb{F}^+$ is *exactly* the set of all scaling factors. In order to do this we will need to start with a point O and $x \in \mathbb{F}^+$. Then, for each point P, we will need to specify a choice of a point xP in $\overrightarrow{OP}$, as we did with $x = 1$, and prove that $d(O, xP) = xd(O, P)$.

A dilation with center O and scaling factor x magnifies or reduces the distance of each point from O by a factor of x. In Section 3.3 we will prove the Scaling Theorem 78 which says that the same dilation magnifies or reduces the distance between *every* two points by a factor of x. In Section 3.4 we will show that similarity of triangles has two very nice additional characterizations by proving the Similar Triangles Theorem 88.

3.1 The Euclidean Parallel Axiom

In order to prove the Scaling Lemma 69 and Scaling Theorem 78 we will need to add one last axiom and investigate its consequences. This axiom builds on the Weak Parallel Line Corollary 39 to say that parallel lines behave as illustrated in Figure 2.1(ii).

(EUC) Euclidean Parallel Axiom. *For every line ℓ and every point P not on ℓ, there is at most one line containing P that is parallel to ℓ.*

Corollary 53 ([2] C59). *Two lines parallel to the same line are parallel to each other.*

Strong Alternate Interior Angle Theorem 54 ([2] T60). *Assume that a transversal line intersects two lines. Then the two lines are parallel if and only if the alternate interior angles are congruent.*

Corollary 55 ([2] C61). *Assume a transversal line intersects two lines. Then the two lines are parallel if and only if the corresponding angles are congruent.*

If A, B, C and D are four points and the four segments AB, BC, CD and DA have no points of intersection other than those four points, then the union of those four segments is called a **quadrilateral** and is denoted by $\square ABCD$. We say the segments AB and CD are **parallel** if the lines $\overleftrightarrow{AB}$ and $\overleftrightarrow{CD}$ are parallel. A quadrilateral with a pair of opposite sides that are parallel is called a **trapezoid**. A **parallelogram** is a trapezoid in which both pairs of opposite sides are parallel. A **rectangle** is a quadrilateral in which each angle is a right angle. A **rhombus** is a quadrilateral in which all four sides are congruent.

Lemma 56. *Every rectangle is a parallelogram.*

Theorem 57 ([2] T72). *Each pair of opposite sides of a parallelogram are congruent.*

Corollary 58 ([2] C62). *Opposite sides of a rectangle are congruent.*

By the **distance** from a point P to a line ℓ not containing P, we mean the length $\mathcal{L}(PQ)$ where Q is the unique point of ℓ for which $\overleftrightarrow{PQ}$ is perpendicular to ℓ. From Theorem 52 we know that PQ is the shortest segment connecting P to a point on ℓ.

Theorem 59 ([2] T63). *Assume that ℓ and m are two parallel lines. Then all points on m are the same distance from ℓ.*

If m and ℓ are parallel lines, then Theorem 59 allows us to define the **distance** between m and ℓ to be the distance from any point of one of these lines to the other line.

Theorem 60 ([2] T64). *Let $\square ABCD$ be a quadrilateral in which $AB \cong CD$ and $\angle A$ and $\angle D$ are right angles. Then $\square ABCD$ is a rectangle.*

Corollary 61. *Given two segments, there exists a rectangle with one pair of opposite sides congruent to the first segment and the other pair congruent to the second segment.*

A **square** is a rectangle that is also a rhombus, i.e., a quadrilateral with four right angles and four congruent sides.

Theorem 62 ([2] T65). *Given two points A and D, there are points B and C such that $\square ABCD$ is a square.*

A **unit square** is a square, each of whose sides has length one. It follows from Theorem 62 that unit squares exist. They will provide a unit of measure for calculating areas.

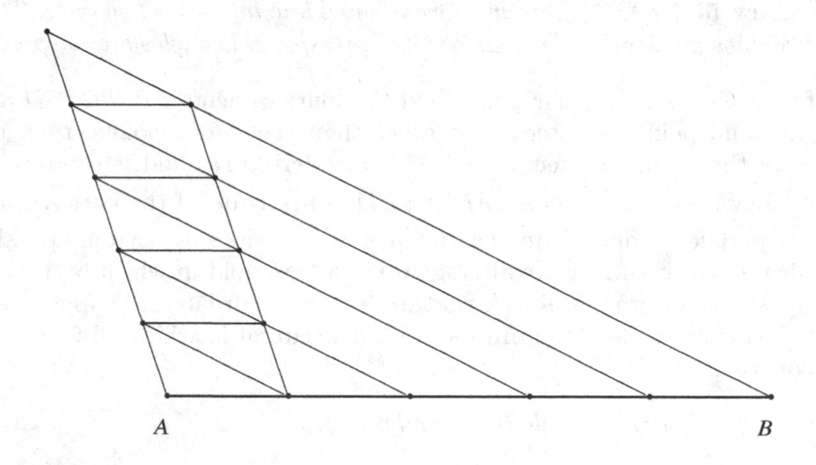

Fig. 3.3 Dividing AB into five congruent segments. [2]

Theorem 63 ([2] T80). *If n is an integer greater than 1, then every segment contains $n-1$ points that divide it into n congruent segments.*

(See Figure 3.3.)

As we saw with Theorem 17, our definition of *congruence* opens the possibility of proving many figures congruent. Here is another example.

Theorem 64. *Two rectangles are congruent if and only if their corresponding sides are congruent.*

(Consider the rectangles $\square ABCD$ and $\square A'B'C'D'$ in Figure 3.4 with congruent corresponding sides. By SAS there is an isometry α taking A, B and C to A', B' and C', respectively. Show that $D^\alpha = D'$.)

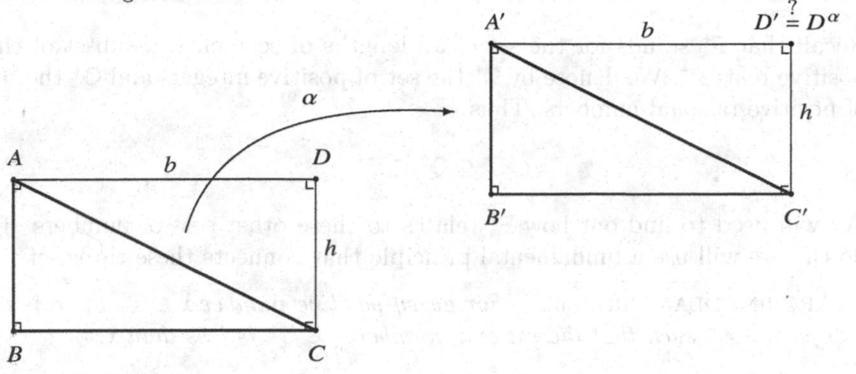

Fig. 3.4 $\square ABCD \cong \square A'B'C'D'$

3.2 The Scaling Lemma

In this section we will show that every number in $\mathbb{F}^+$ is a scaling factor. Let O be a fixed point. We will start by identifying, for each $x \in \mathbb{F}^+$ and each point P, a point we will call xP. Our goal will be to prove that $P \mapsto xP$ is a dilation with center O and scaling factor x.

We begin by taking $xO := O$. For $P \neq O$ let I be either one of the two points on the line through O perpendicular to $\overleftrightarrow{OP}$ at distance one from O. Let xI denote the unique point on $\overrightarrow{OI}$ at distance x from O. By Theorems 26 and 35 there exists a unique ray $\overrightarrow{xI\ M}$ on the P-side of $\overleftrightarrow{OI}$ so that $\angle O\ xI\ M \cong \angle OIP$. (See Figure 3.5.)

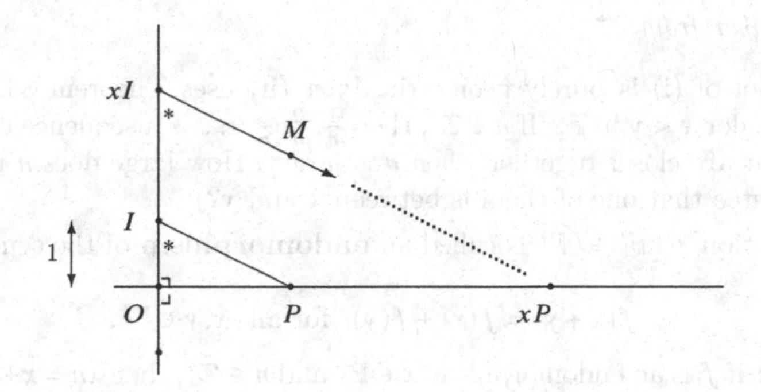

Fig. 3.5 Definition of xP.

Lemma 65. *The ray $\overrightarrow{xI\ M}$ intersects the ray $\overrightarrow{OP}$ at a point that we will denote by xP.*

The Scaling Lemma 69 will tell us that $\mathcal{L}(O\ xP) = x\mathcal{L}(OP)$. Proving this lemma will require some facts about numbers and an idea from algebra.

Recall that $\mathbb{F}^+$ stands for the set of all lengths of segments, a subset of the positive reals $\mathbb{R}^+$. We denote by $\mathbb{Z}^+$ the set of positive integers and $\mathbb{Q}^+$ the set of positive rational numbers. Thus

$$\mathbb{Z}^+ \subseteq \mathbb{Q}^+ \subseteq \mathbb{R}^+.$$

We will need to find out how $\mathbb{F}^+$ relates to these other sets of numbers. To do this we will use a fundamental principle that connects these three sets.

ARCHIMEDEAN PROPERTY. *For every positive number $x \in \mathbb{R}^+$ there is an integer $n \in \mathbb{Z}^+$ such that the rational number $\frac{1}{n} \in \mathbb{Q}^+$ is less than x.*

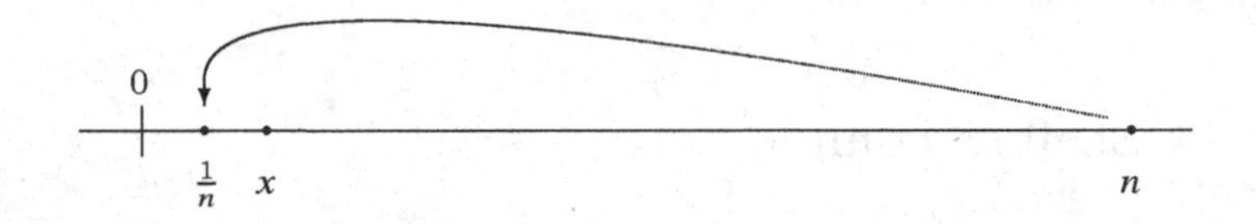

Fig. 3.6 The Archimedean Property.

First $\mathbb{F}^+$ Lemma 66.

(i) *The set $\mathbb{F}^+$ is closed under addition, that is, $x + y \in \mathbb{F}^+$ whenever x and y are in $\mathbb{F}^+$. Consequently $\mathbb{F}^+$ forms a sub-semigroup of the additive semigroup $\mathbb{R}^+$.*

(ii) $\mathbb{Q}^+ \subseteq \mathbb{F}^+$

(iii) *$\mathbb{Q}^+$ is dense in $\mathbb{F}^+$, that is, between any two numbers in $\mathbb{F}^+$ there is a number from $\mathbb{Q}^+$.*

(The proof of (i) is purely geometric. Part (ii) uses Theorem 63. To prove (iii), consider $x < y$ in $\mathbb{F}^+$. If $n \in \mathbb{Z}^+$, then $\frac{1}{n}, \frac{2}{n}, \frac{3}{n}, \ldots$ is a sequence of numbers in $\mathbb{Q}^+$ that are closer together when n is larger. How large does n need to be to guarantee that one of them is between x and y?)

A function $f : \mathbb{F}^+ \to \mathbb{F}^+$ is called an **endomorphism** of the semigroup $\mathbb{F}^+$ if

$$f(x + y) = f(x) + f(y) \quad \text{for all} \quad x, y \in \mathbb{F}^+.$$

Note that if f is an endomorphism, $x \in \mathbb{F}^+$ and $n \in \mathbb{Z}^+$, then $xn = x+x+\cdots+x \in \mathbb{F}^+$ and

$$f(xn) = f(x + x + \cdots + x) = f(x) + f(x) + \cdots + f(x) = f(x)n.$$

Our next lemma gives a characterization of the endomorphisms of $\mathbb{F}^+$.

Endomorphism Lemma 67. *If f is a function from $\mathbb{F}^+$ to $\mathbb{F}^+$, then it is an endomorphism of $\mathbb{F}^+$ if and only if there is an $a \in \mathbb{F}^+$ such that $f(x) = xa$ for all $x \in \mathbb{F}^+$.*

(Let f be an endomorphism of $\mathbb{F}^+$ and let $a := f(1)$. You need to show that $f(x) = xa$ for all $x \in \mathbb{F}^+$. First prove this for $x \in \mathbb{Z}^+$, then for $x = \frac{1}{n}$ with $n \in \mathbb{Z}^+$ and then for $x = \frac{m}{n} \in \mathbb{Q}^+$. Next prove that f is increasing, that is, $x < y$ if and only if $f(x) < f(y)$. To see that $f(x) = xa$ for $x \in \mathbb{F}^+$, show that the same rational numbers are less than x as are less than $f(x)/a$. Then apply the 1st $\mathbb{F}^+$ Lemma 66(iii).)

Lemma 68. *Let P and I be two points such that OI is a unit interval and $\overleftrightarrow{OI} \perp \overleftrightarrow{OP}$. Then the function $f : \mathbb{F}^+ \to \mathbb{F}^+$ defined as*

$$f(x) := d(O, xP)$$

is an endomorphism of the additive semigroup $\mathbb{F}^+$.

(See Figure 3.7.)

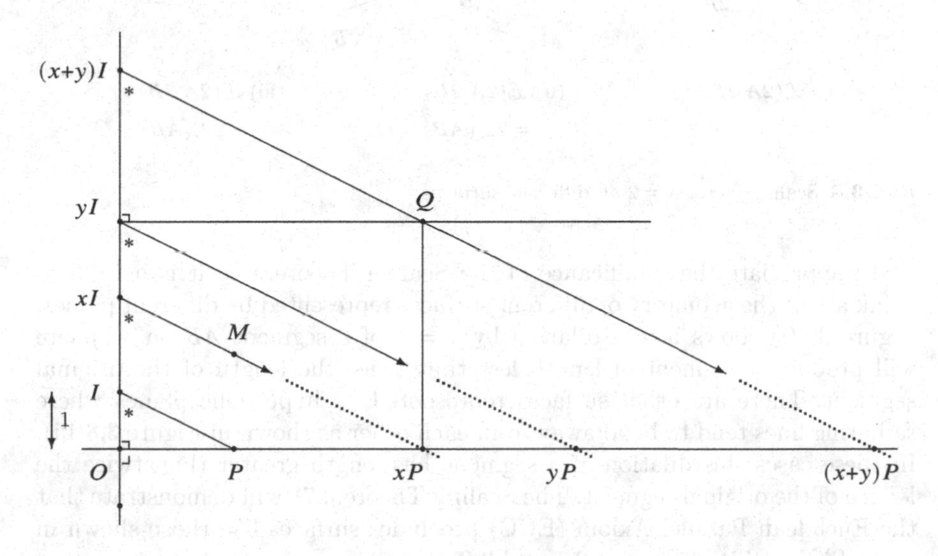

Fig. 3.7 The function f is an endomorphism.

Scaling Lemma 69. *Let $x > 0$ be a real number and let O be a point.*

(i) *If $x \notin \mathbb{F}^+$, then x is not a scaling factor.*
(ii) *If $x \in \mathbb{F}^+$, then $P \mapsto xP$ is a dilation with center O and scaling factor x. In particular, each $x \in \mathbb{F}^+$ is a scaling factor.*

2nd $\mathbb{F}^+$ Lemma 70. *The set $\mathbb{F}^+$ is closed under multiplication, inversion and division, that is, xy and y^{-1} and x/y are in $\mathbb{F}^+$ whenever x and y are in $\mathbb{F}^+$.*

3.3 The Scaling Theorem

A dilation α with center O and scaling factor x dilates the distance between O and any other point by a factor of x. In this section we will prove that α dilates the distance between *every* two points by a factor of x. This theorem is illustrated in Figure 3.8(ii) with scaling factor $x = 2$.

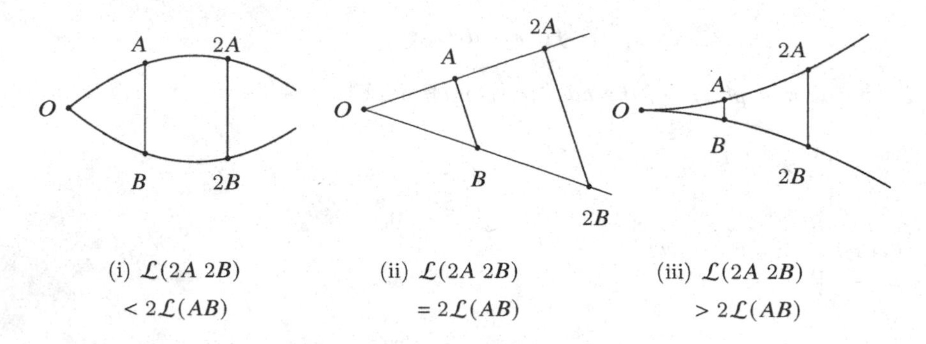

Fig. 3.8 Scaling factor $x = 2$ on different surfaces.

To appreciate the significance of the Scaling Theorem 78 it is helpful to think about the geometry of different surfaces represented by different planes. Figure 3.8(i) shows how a dilation by $x = 2$ of a segment AB on a sphere will produce a segment of length less that twice the length of the original segment. There are other surfaces represented by hyperbolic planes where radiating lines tend to bend away from each other as shown in Figure 3.8(iii). In these cases this dilation of a segment has length greater than twice the length of the original segment. The Scaling Theorem 78 will demonstrate that the Euclidean Parallel Axiom (EUC) precludes surfaces like those shown in parts (i) and (iii) of Figures 2.1 and 3.8.

Our first step is to show that the Scaling Theorem is true in the special case that the two points are collinear with O.

Lemma 71. *Let xA and xB be the images of two points A and B under the dilation with center O and scaling factor x. If $O \in \overleftrightarrow{AB}$, then $d(xA, xB) = xd(A, B)$.*

It remains to prove the Scaling Theorem for points A, B and O that are not collinear. Since $2 \in \mathbb{F}^+$, it is a scaling factor by the Scaling Lemma. We start with this simplest case in which the scaling factor is 2. The proofs for all other scaling factors will be progressive extensions of this proof. Since this context is focused on triangles, it is more natural to think in terms of length measure $\mathcal{L}$ instead of distance measure d.

Lemma 72 ([2] T112). *If A, O and B are not collinear, then*

(i) *the length of the segment $2A\ 2B$ is $\mathcal{L}(2A\ 2B) = 2\mathcal{L}(AB)$ and*
(ii) $\overleftrightarrow{2A\ 2B}$ *is parallel to* $\overleftrightarrow{AB}$.

(Construct Figure 3.9 carefully, verifying the congruences shown as you go. Then conclude that $F = 2B$.)

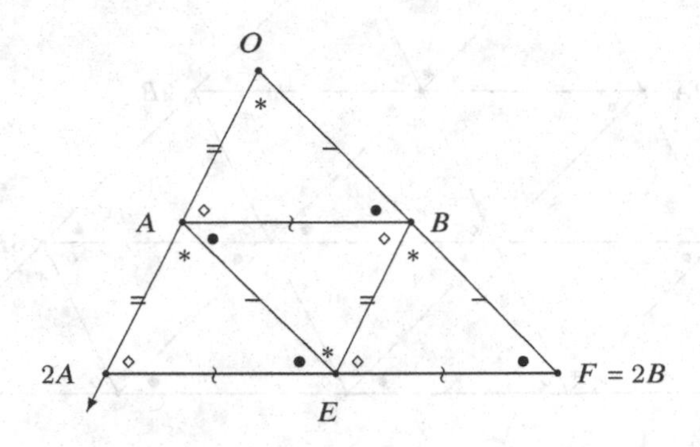

Fig. 3.9 Scaling Theorem for scaling factor 2.

Lemma 73. *If A, O and B are non-collinear and n is a positive integer, then*

(i) *the length of the segment $nA\ nB$ is $\mathcal{L}(nA\ nB) = n\mathcal{L}(AB)$ and*
(ii) $\overleftrightarrow{kA\ kB}$ *is parallel to* $\overleftrightarrow{AB}$ *for $k = 2, 3, \ldots, n$.*

(Prove this lemma as you did the previous one, working down from the top and verifying the labels on the right.)

Since $\mathbb{F}^+$ is exactly the set of scaling factors and is closed under multiplication (Lemmas 69 and 70), we can state the following lemma.

Lemma 74. *Let x and y be scaling factors for dilations with the same center O and let A be a point. Then $x(yA) = (xy)A$.*

Recall from Lemma 66 that the positive rational numbers $\mathbb{Q}^+$ are a subset of $\mathbb{F}^+$. Consequently every positive rational number is a scaling factor by the Scaling Lemma 69.

Lemma 75. *If A, O and B are non-collinear and $r = \frac{m}{n}$ is a positive rational number, then*

(i) *the length of the segment $rA\ rB$ is $\mathcal{L}(rA\ rB) = r\mathcal{L}(AB)$, and*
(ii) *line* $\overleftrightarrow{\frac{k}{n}A\ \frac{k}{n}B}$ *is parallel to* $\overleftrightarrow{AB}$ *if $k \in \mathbb{Z}^+$ and $m \geq k \neq n$.*

Lemma 76. *If A, O and B are non-collinear and $x \neq 1$ is a scaling factor, then* $\overleftrightarrow{xA\ xB}$ *is parallel to* $\overleftrightarrow{AB}$.

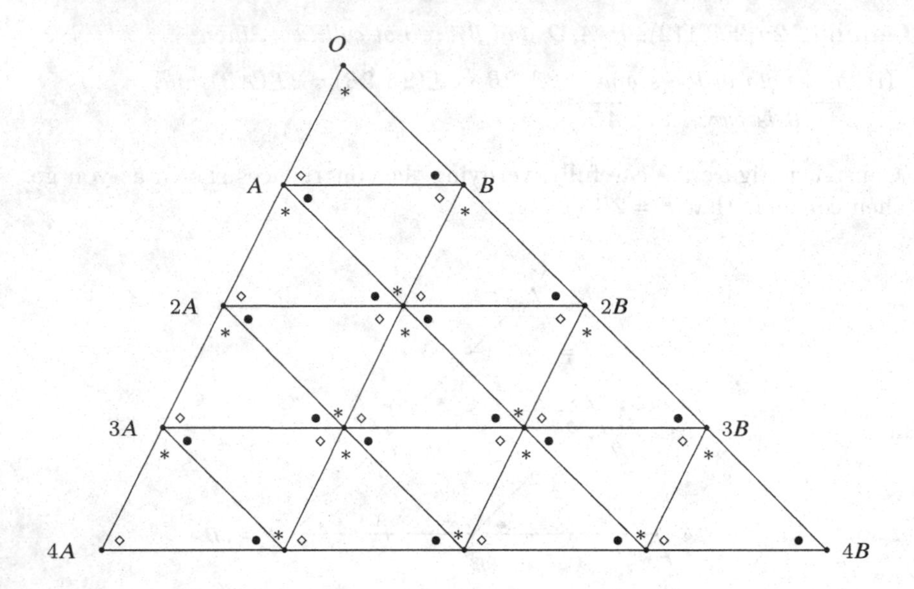

Fig. 3.10 Scaling Theorem for scaling factor $n = 4$.

(Suppose x is irrational and $\overleftrightarrow{xA\ xB}$ is not parallel to $\overleftrightarrow{AB}$. Let $C \neq xA$ be the point on $\overrightarrow{OA}$ for which $\overleftrightarrow{C\ xB}$ is parallel to $\overleftrightarrow{AB}$ (Figure 3.11). For each positive integer n there is a non-negative integer m such that $\frac{m}{n} < x < \frac{m+1}{n}$. Show that $[\frac{m}{n}B\ xB\ \frac{m+1}{n}B], [\frac{m}{n}A\ xA\ \frac{m+1}{n}A]$ and $[\frac{m}{n}A\ C\ \frac{m+1}{n}A]$. Then $\mathcal{L}(C\ xA) < \mathcal{L}(\frac{m}{n}A\ \frac{m+1}{n}A)$ for every choice of n. Now show that $\mathcal{L}(\frac{m}{n}A\ \frac{m+1}{n}A) = \frac{1}{n}\mathcal{L}(OA)$. This gives you a contradiction if you choose n large enough so that $\frac{1}{n}\mathcal{L}(OA) < \mathcal{L}(C\ xA)$.)

Lemma 77. *Let A, O and B be non-collinear points and let x and y be scaling factors for dilations centered at O. Then $x < y$ if and only if $\mathcal{L}(xA\ xB) < \mathcal{L}(yA\ yB)$.*

(Use Lemma 76.)

We can now extend Lemma 75(i) to all scaling factors in $\mathbb{F}^+$.

Scaling Theorem 78 ([2] p57). *If A and B are two points and $x \in \mathbb{F}^+$ is the scaling factor for a dilation, then $\mathcal{L}(xA\ xB) = x\mathcal{L}(AB)$.*

(The number $y := \mathcal{L}(xA\ xB)/\mathcal{L}(AB)$ tells us how many times larger or smaller $\mathcal{L}(xA\ xB)$ is than $\mathcal{L}(AB)$. You would like to prove that y is x. Do this by showing that a rational number q is less that y if and only it it is less than x. Then apply Lemma 66(iii).)

The Scaling Theorem 78 has many useful consequences. The first we give is Axiom 8 of [2], which we can now prove.

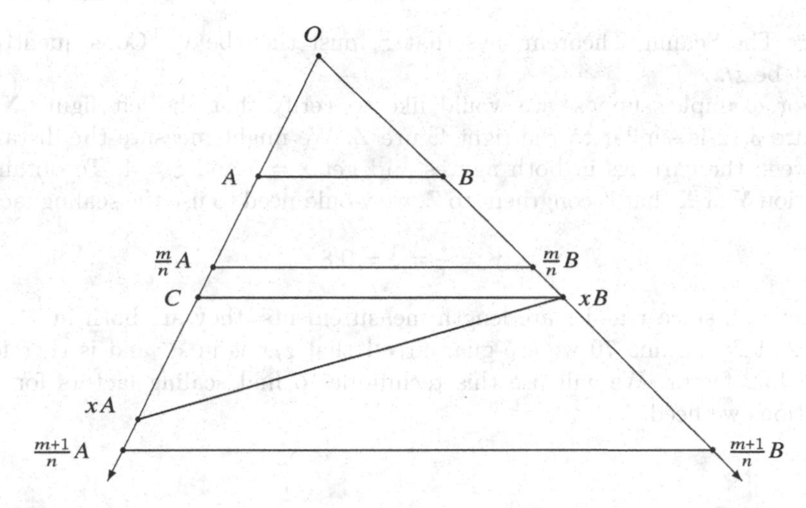

Fig. 3.11 Suppose $\overleftrightarrow{xA\,xB}$ is not parallel to $\overleftrightarrow{AB}$.

Theorem 79 ([2] Axiom 8). *Let A, B and C be three points and let xA, xB and xC be their images under a dilation with center O and scaling factor x.*

(i) *If A, B and C are collinear, then xA, xB and xC are collinear.*
(ii) *If B is between A and C, then xB is between xA and xC.*

Corollary 80 ([2] T110). *Let ℓ be a line and let ℓ' be the image of ℓ under a dilation. Then either $\ell' = \ell$ or ℓ' is a line parallel to ℓ.*

Corollary 81 ([2] T111). *Dilations preserve angle congruence. Specifically, if non-collinear points A, B and C go to non-collinear points xA, xB and xC under a dilation with scaling factor x, then*

$$\angle ABC \cong \angle xA\,xB\,xC.$$

3.4 Similarity

Recall that figures are similar if one is congruent to the image of the other under some dilation. As we now know a great deal about congruence, our primary challenge to showing that two figures **X** and **Z** are similar is that of finding the right scaling factor. Assume the **Y** is congruent to **Z** and **Y** is the image of **X** under a dilation with scaling factor y. The Scaling Theorem suggests a way to determine the value of y. Suppose we measure the distance between two points of **X** and get a value x. Then we measure the distance between the corresponding two points of **Z** and get a value z. Since **Y** is congruent to **Z**, the corresponding measurement on **Y** would also have to be z. The Scaling Theorem says that z must then be xy. Consequently y must be z/x.

For example, suppose we would like to verify that the left figure **X** in Figure 3.12 is similar to the right figure **Z**. We might measure the distance between the earrings in both figures and get $x = 5$ and $z = 4$. To obtain a dilation **Y** of **X** that is congruent to **Z**, we would need to use the scaling factor

$$y := \tfrac{z}{x} = \tfrac{4}{5} = 0.8.$$

In general, since x and z are length measurements, they are both in $\mathbb{F}^+$. By the 2nd $\mathbb{F}^+$ Lemma 70 we are guaranteed that z/x is in $\mathbb{F}^+$ and is therefore a scaling factor. We will use this technique to find scaling factors for the dilations we need.

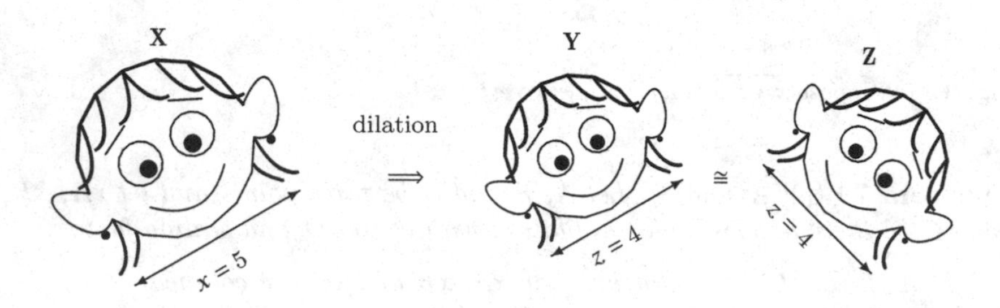

Fig. 3.12 Finding the scaling factor y.

Corollary 82 ([2] P109). *Any two segments are similar.*

Corollary 83 ([2] P106). *Any two circles are similar.*

As applications of the Scaling Theorem we will give two other criteria for *triangles* to be similar, neither of which refer to an isometry or a dilation. We say that two sequences of numbers $a, b, c, \ldots$ and $a', b', c', \ldots$ are **proportional** if there is a number x such that

$$\tfrac{a'}{a} = \tfrac{b'}{b} = \tfrac{c'}{c} \cdots = x.$$

Lemma 84 ([2] T116). *If $\triangle ABC \sim \triangle A'B'C'$, then corresponding angles are congruent and corresponding sides are proportional by the scaling factor of the dilation that was used.*

Theorem 85 ([2] T118). *If a triangle with sides of lengths a, b and c is similar to a triangle with corresponding sides of length a', b' and c', then corresponding pairs of sides are in the same ratio:*

$$\tfrac{a'}{b'} = \tfrac{a}{b}, \quad \tfrac{a'}{c'} = \tfrac{a}{c}, \quad \tfrac{b'}{c'} = \tfrac{b}{c}.$$

We would like to prove two converses for Lemma 84: that congruence of corresponding angles and proportionality of corresponding sides each imply that two triangles are similar. Since proving similarity requires finding the right scaling factor, we will again need to use the closure of $\mathbb{F}^+$ under division (2nd $\mathbb{F}^+$ Lemma 70).

Proportional Sides Lemma 86 ([2] T120). *If the corresponding sides of two triangles are proportional, then the triangles are similar.*

Lemma 87 ([2] T123 **AAA**). *If the corresponding angles of two triangles are congruent, then the triangles are similar.*

Combining Lemmas 84, 86 and 87 gives us a full characterization of similar triangles.

Similar Triangles Theorem 88 ([2] T128). *For triangles* $\mathbf{T}$ *and* $\mathbf{T}'$, *the following are equivalent; that is, each one implies the other two.*

(i) $\mathbf{T}$ *is similar to* $\mathbf{T}'$.
(ii) *Each angle of* $\mathbf{T}$ *is congruent to the corresponding angle of* $\mathbf{T}'$.
(iii) *The sides of* $\mathbf{T}$ *are proportional to the sides of* $\mathbf{T}'$.

Theorem 89 ([2] C126 **AA**). *If two angles of one triangle are congruent to two corresponding angles of another triangle, then the triangles are similar.*

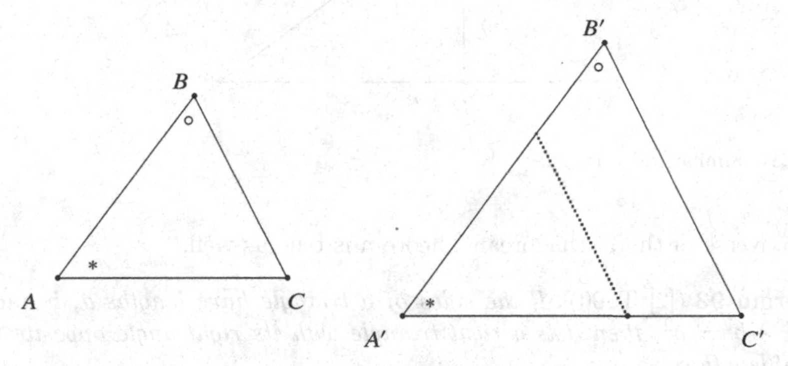

Fig. 3.13 AA.

Corollary 90 ([2] C127 **A**). *If a non-right angle of one right triangle is congruent to a non-right angle of another right triangle, then the triangles are similar.*

Lemma 91 ([2] C129). *Let* $\triangle ABC$ *be a right triangle with right angle at* C *and let* D *be the intersection with* $\overleftrightarrow{AB}$ *of the perpendicular from* C. *Then* $[ADB]$ *and*

$$\triangle ABC \sim \triangle ACD \sim \triangle CBD.$$

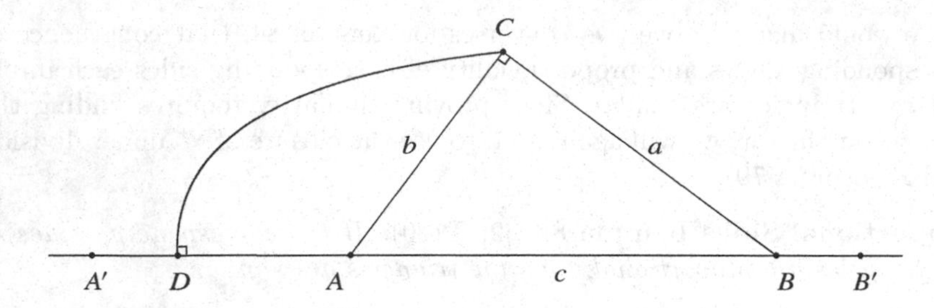

Fig. 3.14 Why must D be between A and B?

As a result of the Similar Triangles Theorem 88 we now have enough background in place to give a full proof of the Pythagorean Theorem.

2nd Pythagorean Theorem 92 ([2] P130). *If a right triangle has legs of lengths a and b and hypotenuse of length c, then $c^2 = a^2 + b^2$.*

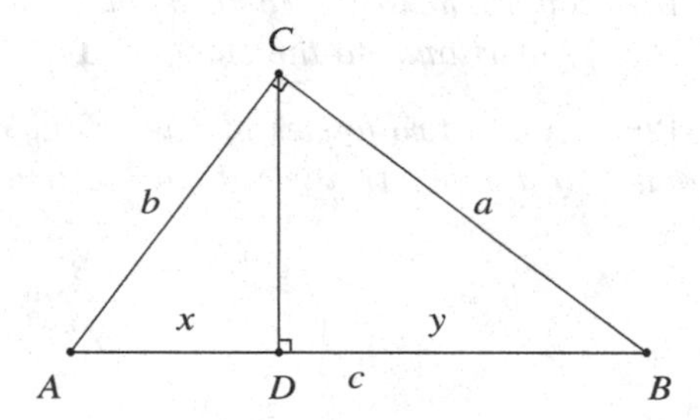

Fig. 3.15 Similar right triangles. [2]

The converse of the Pythagorean Theorem is true as well.

Theorem 93 ([2] T100). *If the sides of a triangle have lengths a, b and c and $a^2 + b^2 = c^2$, then it is a right triangle with its right angle opposite the side of length c.*

Chapter 4
Area Measure

An area measure function $\mathcal{A}$ can be introduced either as another primitive construct or as a defined construct. The text [2] takes $\mathcal{A}$ as a primitive construct governed by an axiom similar to our Area Measure Theorem 120 at the beginning of Section 4.4. This amounts to a shortcut that is available to our readers as well. By assuming the existence of an area measure function $\mathcal{A}$ with the properties listed in Theorem 120, it is possible to go directly from the definition of *polygonal region* on page 38 to Section 4.5 and calculate the polygonal areas as is done in Chapter 3 of [2]. It is also possible to omit most of our Section 7.1.

The price for taking this shortcut is that when the time comes to apply the theorems we prove here to a particular plane, as we do in Chapter 8, it will then be necessary to demonstrate the existence of an area measure function $\mathcal{A}$ satisfying the conditions of Theorem 120 in each of those planes. In order to offer a full axiomatic development of this geometry, we will forego this shortcut.

In Section 4.1 of this chapter we will identify a set of figures in the plane $\mathcal{P}$, called *polygonal regions*, whose areas we would initially like to define. In Section 4.2 we will present the historic forerunner of area measure used by the early Greeks. In Section 4.3 we will present the definition of an area measure function on all polygonal regions. Our starting point for this definition is our Lemma 102, which uses the Similar Triangles Theorem 88 to define triangular area. It is for this reason that we present similarity before area measure. In Section 4.4 we will prove the Area Measure Theorem 120 holds under this definition of area. In Section 4.5 we will use Theorem 120 to establish the standard area formulas for familiar quadrilaterals.

In Chapter 7 we will continue the topic of area measure. Section 7.1 is primarily devoted to a proof of Theorem 199, which we will use to considerably expand the class of regions whose areas we can compute. The work in this section is largely topological in nature, but our presentation is fully self contained.

D. M. Clark, S. Pathania, *A Full Axiomatic Development of High School Geometry*,
https://doi.org/10.1007/978-3-031-23525-2_4

4.1 Polygonal Regions

Recall that "figure" or "region" refers to any set of points. In this section we will identify the regions **F** whose areas we will calculate in this chapter and examine their relevant properties. We begin by looking at parts and characteristics of regions in general. By an **open disc** we mean the inside of some circle. Let **F** be an arbitrary region. The **boundary** of **F** is the set $\mathbf{F}^b$ of all points P in **F** such that every open disc containing P intersects the complement of **F**. The **interior** of **F** is the set $\mathbf{F}^i$ of all points that are in an open disc entirely contained in **F**. Thus every region **F** is the disjoint union

$$\mathbf{F} = \mathbf{F}^b \cup \mathbf{F}^i$$

of its boundary and its interior.

Triangles will play a central role in area measure. The **inside** of a triangle is the intersection of the interiors of its angles. A **triangular region** is the union of a triangle and its inside. A **triangulation** of a region is a finite set of triangular regions with pairwise disjoint interiors whose union is that region. A **polygonal region** is a region **F** that has a triangulation. (See Figure 4.1). In this chapter we will define and calculate areas of polygonal regions.

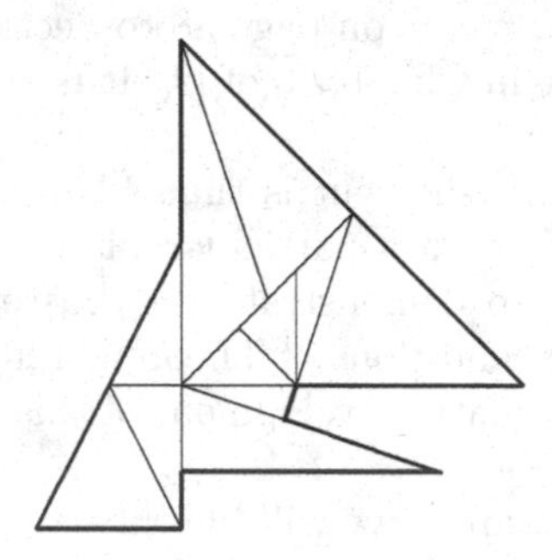

Fig. 4.1 A triangulation of a polygonal region **F**.

To do this we will start by defining a particular class of familiar figures which we will prove to be polygonal regions and which have boundaries and interiors that are easily identified. Let $n \geq 3$ be an integer. A figure **X** is called a **convex** n**-gon** if there is a sequence of n points $P_0, P_1, \ldots, P_{n-1}$ (called the **vertices** of **X**) forming n segments $P_0P_1, P_1P_2, \ldots, P_{n-2}P_{n-1}, P_{n-1}P_0$ (called the **sides** of **X**) such that, if AB is a side, then the $n-2$ vertices other than A and B are on the same side of $\overleftrightarrow{AB}$ and

$$\mathbf{X} := P_0P_1 \cup P_1P_2 \cup \ldots P_{n-2}P_{n-1} \cup P_{n-1}P_0.$$

The n angles $\angle P_iP_{i+1}P_{i+2}$ for $i < n-2$, together with $\angle P_{n-2}P_{n-1}P_0$ and $\angle P_{n-1}P_0P_1$, are called the **angles** of **X**. We write

$$\mathbf{X} := \bigcirc P_0 P_1 P_2 \ldots P_{n-1}$$

if **X** forms a convex n-gon when the vertices are listed in this order. A figure is a **convex polygon** if it is a convex n-gon for some $n \geq 3$. For example, a figure is a triangle if and only if it is a convex 3-gon. Every parallelogram is a convex 4-gon.

In this context the term *convex* has a very general meaning that we will also need. A region **F** is **convex** if it has the property that

$$A, C \in \mathbf{F} \text{ and } [ABC] \text{ implies } B \in \mathbf{F}.$$

For example, it is straightforward to prove from facts in Section 1.1 that many regions we have discussed are convex.

Lemma 94. *Every line, each side of a line and each segment is convex.*

However, convex polygons are not convex in this sense. We call them *convex polygons* because, as we will soon establish, they serve to enclose a region that is convex.

Relative to a convex polygon **X** with side AB, we call the side of $\overleftrightarrow{AB}$ containing the other vertices the **inside** of $\overleftrightarrow{AB}$ and the other side of $\overleftrightarrow{AB}$ the **outside** of $\overleftrightarrow{AB}$. The **inside** of **X**, denoted by $\text{In}(\mathbf{X})$, is the intersection of the insides of all the sides of **X**. Equivalently, it is the intersection of the interiors of the angles of **X**. We refer to the set

$$\overline{\mathbf{X}} := \mathbf{X} \cup \text{In}(\mathbf{X})$$

as a **convex polygonal region**. Note that this is a disjoint union because no point of $\text{In}(\mathbf{X})$ can be on a side of **X**.

Convex polygonal regions prove to be very useful, in part because we can show that they have interiors and boundaries that are easily recognized. To do this we need a lemma giving a basic property of open discs that we will need here and will make extensive use of in Chapter 7.

Lemma 95. *If the point P is in the open disc $\mathbf{D}_1$, then there is an open disc $\mathbf{D}_2 \subseteq \mathbf{D}_1$ centered at P.*

Theorem 96. *Let* **X** *be a convex polygon.*

(i) $\overline{\mathbf{X}}^i = \text{In}(\mathbf{X})$: *the interior of* $\overline{\mathbf{X}}$ *is the inside of* **X**.
(ii) $\overline{\mathbf{X}}^b = \mathbf{X}$: *the boundary of* $\overline{\mathbf{X}}$ *is* **X**.

Our goal in the remainder of this section is to justify our terminology by proving that convex polygonal regions are indeed both convex regions and polygonal regions.

Lemma 97. *Every convex polygonal region* $\overline{\mathbf{X}}$ *is a convex region.*

To prove that a convex polygonal region $\overline{\mathbf{X}} = \overline{\bigcirc P_0P_1P_2 \ldots P_{n-1}}$ is a polygonal region we will prove that the diagonal segments from P_0 to each of the other vertices form a triangulation of $\overline{\mathbf{X}}$ as shown in Figure 4.2.

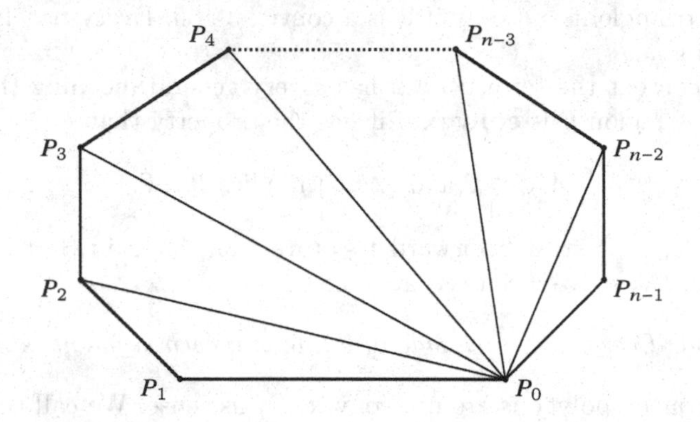

Fig. 4.2 Triangulation of $\overline{\mathbf{X}}$.

Lemma 98. *If $n \geq 3$ and $\mathbf{X} = \bigcirc P_0P_1P_2 \ldots P_{n-1}$ is a convex n-gon, then the $n-2$ triangular regions*

$$\{\overline{\triangle P_0P_iP_{i+1}} \mid 1 \leq i \leq n-2\}$$

have disjoint interiors.

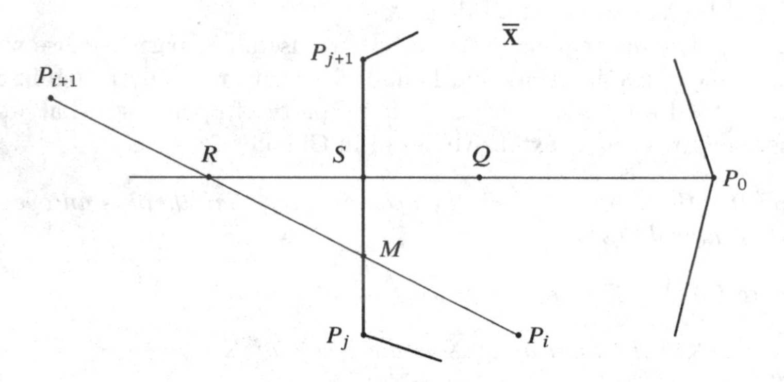

Fig. 4.3 Proof of Lemma 98.

(Suppose that $Q \in \text{In}(\triangle P_0P_iP_{i+1}) \cap \text{In}(\triangle P_0P_jP_{j+1})$ where we have $1 \leq i < j \leq n-2$. By the Crossbar Theorem 11 there are points R and S on $\overrightarrow{P_0Q}$ with $[P_iRP_{i+1}]$ and $[P_jSP_{j+1}]$. Now follow Figure 4.3 to reach a contradiction.)

Theorem 99. *If $n \geq 3$ and $\mathbf{X} = \bigcirc P_0P_1P_2\ldots P_{n-1}$ is a convex n-gon, then $\overline{\mathbf{X}}$ is the union of the $n-2$ triangular regions*

$$\overline{\mathbf{X}} = \bigcup\{\overline{\triangle P_0P_iP_{i+1}} \mid 1 \leq i \leq n-2\}$$

which have disjoint interiors, and is therefore a polygonal region.

(Let $\mathbf{Z}$ denote the union on the right and let $Q \in \overline{\mathbf{X}}$. If Q is on one of these triangles, then it is in $\mathbf{Z}$. If it is not, show that it is in the interior of some triangle $\triangle P_0P_iP_{i+1}$ as shown in Figure 4.4.)

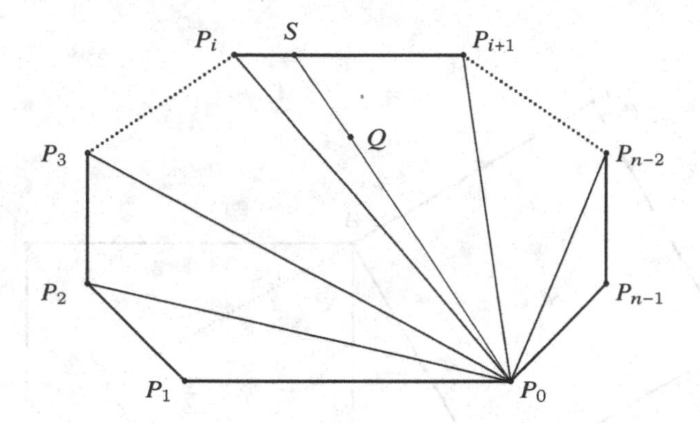

Fig. 4.4 Proof of Theorem 99.

4.2 Decompositional Equivalence

Before we discuss area measure, it is helpful to digress briefly to look at the forerunner of this concept that is best illustrated by the original history of the Pythagorean Theorem. The fact is that our Pythagorean Theorem 92, as we have stated it, was of little interest to the early Greeks. The problem was that the only numbers available at that time were the rational numbers. The Pythagoreans knew, for example, that the only positive integers a and b for which $a^2 + b^2$ was the square of a (rational) number were those rare cases where it was the square of an integer. In all other cases the hypotenuse of a right triangle with legs of length a and b had no numerical length that they knew of, so Theorem 92 simply did not apply.

Around 300 BCE Euclid of Alexandria assembled plane geometry into a single axiomatic development for the first time [4]. His proof of Pythagoras' theorem used the additivity property of angle measure that we will prove in Chapter 5. However, he avoided numerical measurement of areas by using a restricted notion of area equivalence to state and prove Pythagoras' theorem without reference to numbers. We will now present a variant of Euclid's proof that illustrates this technique and avoids mention of either area measure or angle measure.

We say that polygonal regions **F** and **G** are **decompositionally equivalent**, written $\mathbf{F} \overset{de}{\sim} \mathbf{G}$, if there is a positive integer n and sets $\{\mathbf{F}_1, \mathbf{F}_2, \ldots, \mathbf{F}_n\}$ and $\{\mathbf{G}_1, \mathbf{G}_2, \ldots, \mathbf{G}_n\}$ of polygonal regions with disjoint interiors so that

1. $\mathbf{F} = \mathbf{F}_1 \cup \mathbf{F}_2 \cup \cdots \cup \mathbf{F}_n$,
2. $\mathbf{G} = \mathbf{G}_1 \cup \mathbf{G}_2 \cup \cdots \cup \mathbf{G}_n$ and
3. $\mathbf{F}_i \cong \mathbf{G}_i$ for $i = 1, 2, \ldots, n$.

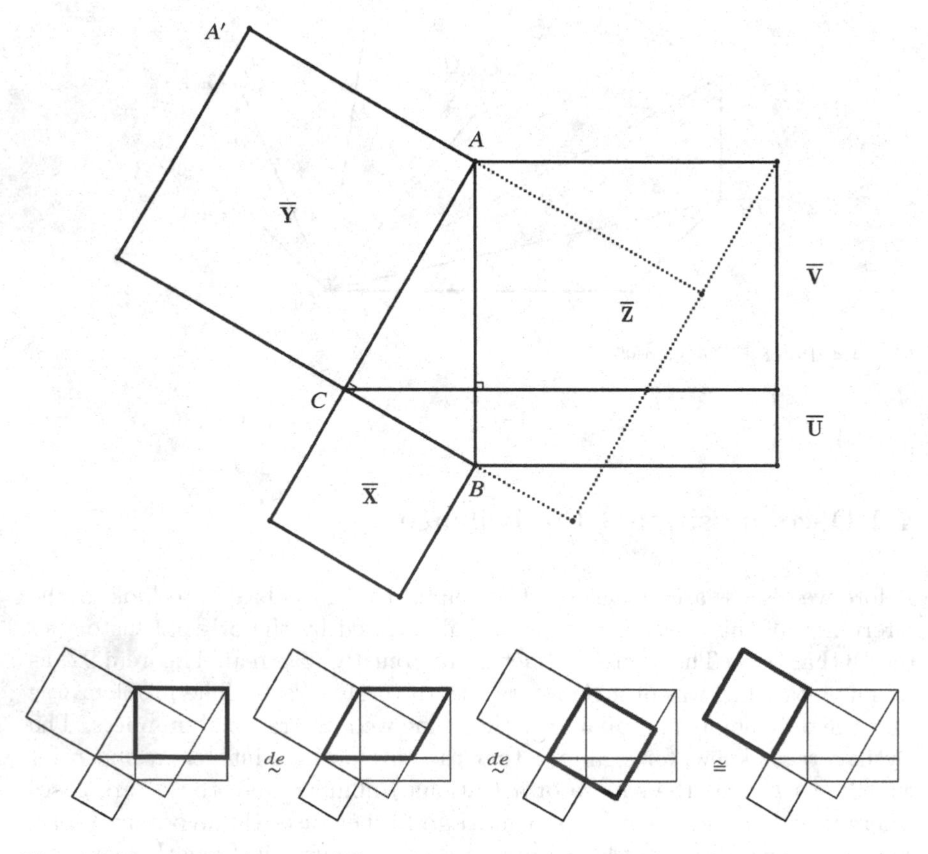

Fig. 4.5 Proof of Pythagorean Theorem 101.

Lemma 100. *Let $\square ABCD$ be a rectangle and let $\square ABEF$ be a parallelogram with $E, F \in \overleftrightarrow{CD}$. Then $\square ABCD \overset{de}{\sim} \square ABEF$.*

3rd Pythagorean Theorem 101. *Given a right triangle, the squares upon its two legs are together decompositionally equivalent to the square upon its hypotenuse.*

(Let $\triangle ABC$ be a right triangle with right angle at C, let $\overline{\mathbf{X}}$, $\overline{\mathbf{Y}}$ and $\overline{\mathbf{Z}}$ be the square regions on the outside of each of the sides of $\triangle ABC$ as shown in Figure 4.5. Then the line perpendicular to $\overleftrightarrow{AB}$ through C divides $\overline{\mathbf{Z}}$ into two rectangular regions $\overline{\mathbf{U}}$ and $\overline{\mathbf{V}}$. Show that $\overline{\mathbf{X}} \overset{de}{\sim} \overline{\mathbf{U}}$ and $\overline{\mathbf{Y}} \overset{de}{\sim} \overline{\mathbf{V}}$.)

Euclid's proof of this theorem is similar to ours but draws on his own tools. Our proof above is possible here because it uses neither an area measure function nor an angle measure function. At the end of this chapter we will use area measure to obtain a different proof of the Pythagorean Theorem 92 as an immediate consequence of Theorem 101 (the Pythagorean Theorem 126).

4.3 Polygonal Area

In this section we will define an area measure function $\mathcal{A}$ on polygonal regions. Ideally we would like to define the area of a polygonal region as the sum of the areas of its component triangular regions, where we *define* the area of a triangular region as half of its base times height. Doing this presents us with two immediate tasks. First we need to establish that calculating base times height gives the same value for all three choices of the base and height. Then we need to show that summing the areas of triangulations gives the same value for all possible triangulations. The first is an immediate consequence of the Similar Triangles Theorem 88.

Lemma 102. *For a triangle $\triangle XYZ$, the value of $bh/2$ is the same for each of the three choices of a base b and corresponding height h.*

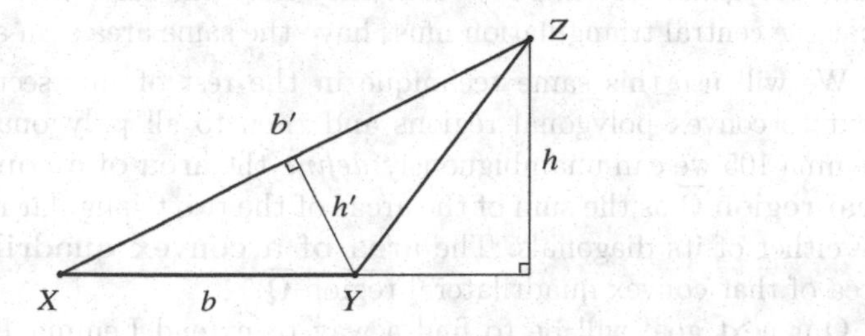

Fig. 4.6 Is $\frac{bh}{2} = \frac{b'h'}{2}$?

Using Lemma 102 we can unambiguously *define* the **area of a triangular region $\overline{\mathbf{T}}$** with a base b and corresponding height h as

$$\mathcal{A}(\overline{\mathbf{T}}) := \frac{bh}{2}.$$

Following standard terminology, the **area of a triangle T** is the area of the triangular region $\overline{\mathbf{T}}$.

The second task will require examining a series of progressively more general cases.

Lemma 103. *If* $\triangle ABC$ *is a triangle with point* P *between* B *and* C*, then*

$$\mathcal{A}(\triangle ABC) = \mathcal{A}(\triangle ABP) + \mathcal{A}(\triangle PBC).$$

Lemma 104. *Opposite vertices of a convex quadrilateral are on opposite sides of the line through the other two vertices.*

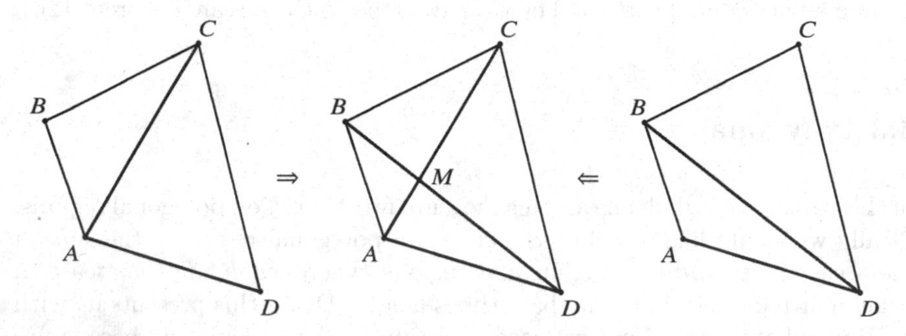

Fig. 4.7 Diagonal triangulations of a convex quadrilateral.

Lemma 105. *If* $\mathbf{Q} = \square ABCD$ *is a convex quadrilateral, then*

$$\mathcal{A}(\triangle ABC) + \mathcal{A}(\triangle ADC) = \mathcal{A}(\triangle BAD) + \mathcal{A}(\triangle BCD).$$

(In Figure 4.7 we have on the left and right two different triangulations of the same polygonal region. Prove that they have the same area sum by showing that the central triangulation must have the same area sum as each of them.)

We will use this same technique in the rest of this section, applying it next to convex polygonal regions and then to all polygonal regions. Using Lemma 105 we can unambiguously *define* the **area of a convex quadrilateral region** $\overline{\mathbf{Q}}$ as the sum of the areas of the two triangular regions produced by either of its diagonals. The **area of a convex quadrilateral Q** is the area of that convex quadrilateral region $\overline{\mathbf{Q}}$.

Our next goal will be to find a way to extend Lemma 105 to all convex polygonal regions.

Lemma 106. *Let* $\triangle ABC$ *be a triangle with points* S *and* R *such that* $[ASB]$ *and* $[BRC]$*. Then*

$$\mathcal{A}(\triangle ABC) = \mathcal{A}(\square ASRC) + \mathcal{A}(\triangle SBR).$$

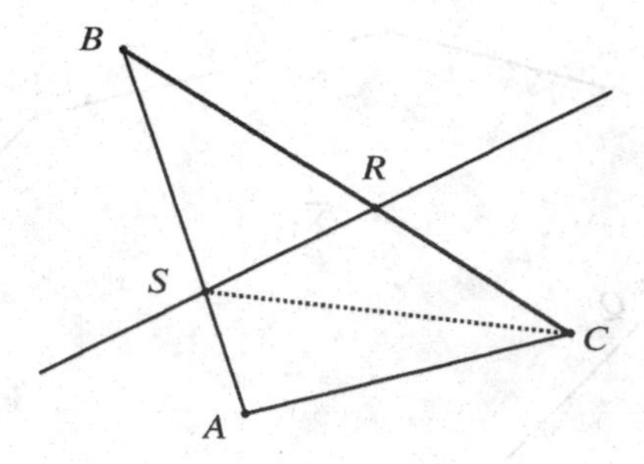

Fig. 4.8 $\mathcal{A}(\triangle ABC) = \mathcal{A}(\square ASRC) + \mathcal{A}(\triangle SBR)$.

Lemma 107. *If a ray $\overrightarrow{r}$ emanates from a point Q inside a convex n-gon* **X**, *then it intersects* **X** *in exactly one point.*

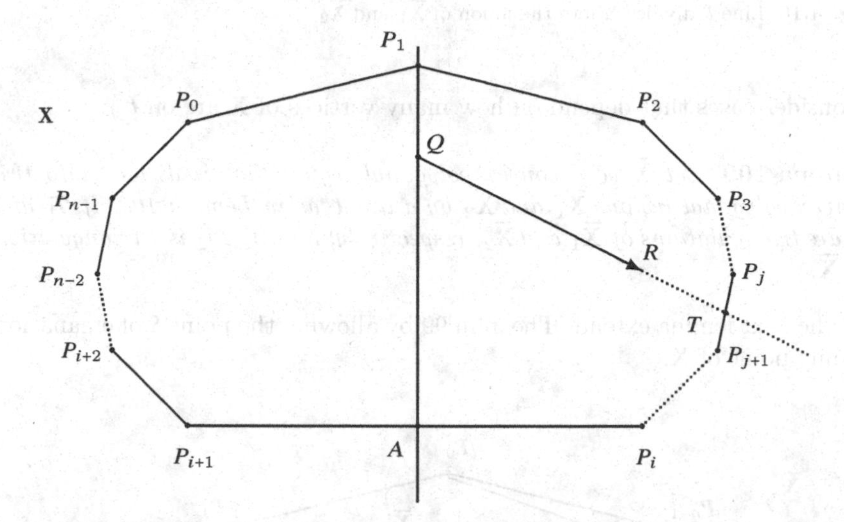

Fig. 4.9 A ray $\vec{r}$ from inside convex polygon **X**.

By a **half plane** of a line ℓ we mean the union of ℓ and one of the sides of ℓ. If A is a point not on ℓ, then we denote by $\mathbf{H}(\ell, A)$ the half plane of ℓ containing A. In this terminology we can describe the triangular region determined by a triangle $\triangle ABC$ as

$$\overline{\triangle ABC} = \mathbf{H}(\overleftrightarrow{AB}, C) \cap \mathbf{H}(\overleftrightarrow{BC}, A) \cap \mathbf{H}(\overleftrightarrow{CA}, B).$$

Lemma 108. *Let* **X** *be a convex polygon and let ℓ be a line containing a point Q inside* **X** *with half planes $\mathbf{H}_1$ and $\mathbf{H}_2$. Then $\overline{\mathbf{X}}_1 := \mathbf{H}_1 \cap \overline{\mathbf{X}}$ and $\overline{\mathbf{X}}_2 := \mathbf{H}_2 \cap \overline{\mathbf{X}}$ are convex polygonal regions with disjoint interiors such that $\overline{\mathbf{X}} = \overline{\mathbf{X}}_1 \cup \overline{\mathbf{X}}_2$.*

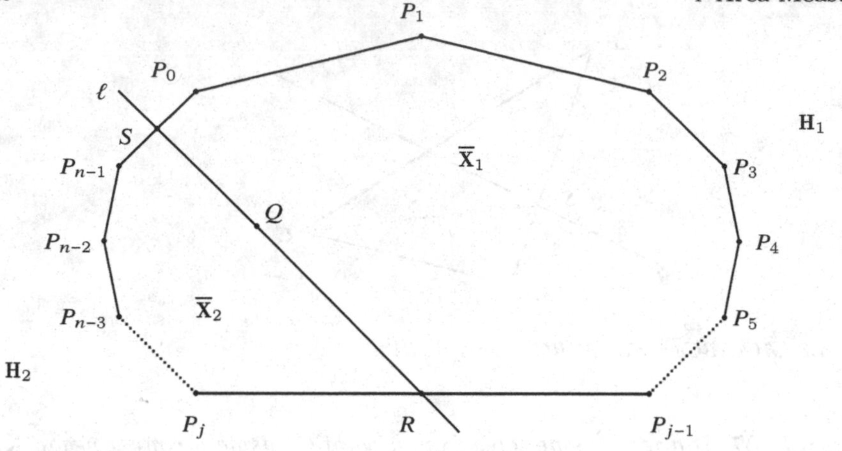

Fig. 4.10 Line ℓ divides $\overline{\mathbf{X}}$ into the union of $\overline{\mathbf{X}}_1$ and $\overline{\mathbf{X}}_2$.

(Consider cases that depend on how many vertices of $\mathbf{X}$ are on ℓ.)

Lemma 109. *Let $\overline{\mathbf{X}}$ be a convex polygonal region that is divided into two convex polygonal regions $\overline{\mathbf{X}}_1$ and $\overline{\mathbf{X}}_2$ by a line ℓ as in Lemma 108. If $\mathcal{T}_1$ and $\mathcal{T}_2$ are triangulations of $\overline{\mathbf{X}}_1$ and $\overline{\mathbf{X}}_2$, respectively, then $\mathcal{T}_1 \cup \mathcal{T}_2$ is a triangulation of $\overline{\mathbf{X}}$.*

The next lemma extends Theorem 99 by allowing the point S of emanation be any point of $\mathbf{X}$.

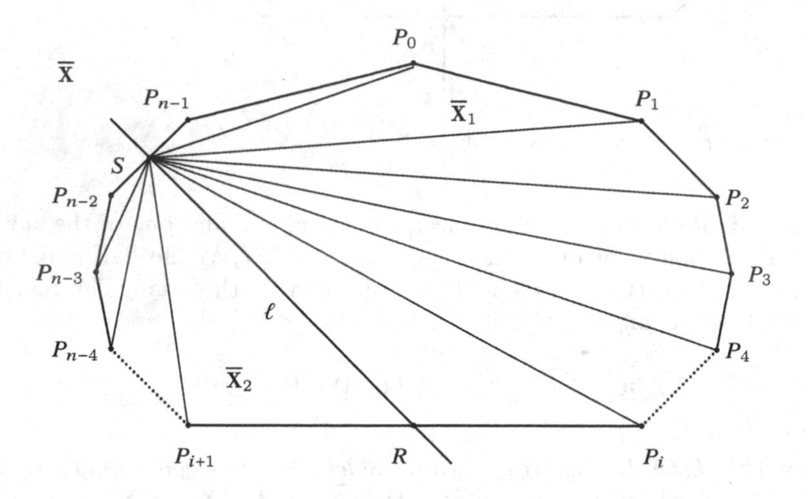

Fig. 4.11 A leaf triangulation of a convex polygonal region.

Lemma 110. *Let $\overline{\mathbf{X}}$ be a convex polygonal region, let $S \in \mathbf{X}$ and let $\mathcal{T}_S$ be the set of triangular regions formed by connecting S to each (other) vertex of $\overline{\mathbf{X}}$. Then $\mathcal{T}_S$ is a triangulation of $\overline{\mathbf{X}}$.*

(See Figure 4.11.)

If $S \in \mathbf{X}$, we will call the triangulation $\mathcal{T}_S$ described in Lemma 110 a **leaf triangulation** of $\overline{\mathbf{X}}$. Building on Lemma 106 we would like to extend the notion of a triangulation to employ other convex polygonal regions as well. Accordingly we say that a **polyation** of a region is a finite set of convex polygonal regions with pairwise disjoint interiors whose union is that region. If we happen to find that a polyation $\mathcal{P}$ consists only of triangular and quadrilateral regions, then we can define its **area sum**, denoted by $\Sigma\mathcal{P}$, as the sum of the areas of those component regions. The next lemma will confront us with just such a polyation. This lemma is a direct extension of Lemma 105 from convex quadrilaterals with diagonal triangulations to all convex polygonal regions and all leaf triangulations.

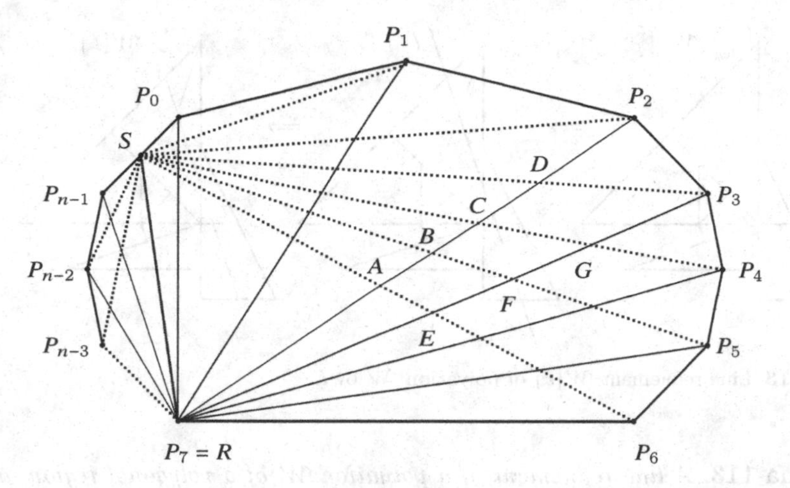

Fig. 4.12 Two leaf triangulations of a convex polygonal region.

Lemma 111. *All leaf triangulations of a convex polygonal region have the same area sum.*

(Referring to Figure 4.12, consider two leaf triangulations radiating from points R and S on a convex polygon $\mathbf{X}$. Show that the two triangulations together form a polyation of $\overline{\mathbf{X}}$ whose convex polygonal regions are all triangular or quadrilateral. Then show that both area sums are the same as the area sum of that polyation.)

Using Lemma 111 we can unambiguously *define* the **area of a convex polygonal region $\overline{\mathbf{X}}$** as the area sum of its leaf triangulations. The **area of a convex polygon $\mathbf{X}$** is the area of the convex polygonal region $\overline{\mathbf{X}}$.

We will now extend the definition of area to all polygonal regions in several steps.

Lemma 112. *Let $\overline{\mathbf{X}}$ be a convex polygonal region that is divided into two convex polygonal regions $\overline{\mathbf{X}}_1$ and $\overline{\mathbf{X}}_2$ by a line ℓ. Then*

$$\mathcal{A}(\overline{\mathbf{X}}) = \mathcal{A}(\overline{\mathbf{X}}_1) + \mathcal{A}(\overline{\mathbf{X}}_2).$$

Consider now a polyation $\mathcal{W}$ of a polygonal region $\mathbf{F}$ and any line ℓ. (Figure 4.13.) We define a new polyation $\mathcal{W}[\ell]$ of $\mathbf{F}$ as follows. If ℓ does not contain a point in the interior of any region of $\mathcal{W}$, take $\mathcal{W}[\ell] := \mathcal{W}$. Otherwise, for each convex polygonal region $\overline{\mathbf{Y}}$ of $\mathcal{W}$ containing a point of ℓ in its interior, apply Lemmas 108 and 112 to replace $\overline{\mathbf{Y}}$ with two convex polygonal regions with disjoint interiors whose union is $\overline{\mathbf{Y}}$ and whose area sums added together give the area of $\overline{\mathbf{Y}}$. As a result, it will follow that $\Sigma\mathcal{W}[\ell] = \Sigma\mathcal{W}$. We refer to $\mathcal{W}[\ell]$ as the **line refinement** of polyation $\mathcal{W}$ by the line ℓ.

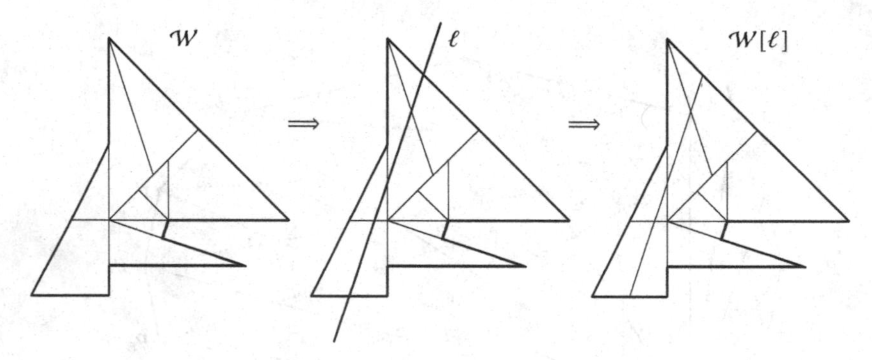

Fig. 4.13 Line refinement $\mathcal{W}[\ell]$ of polyation $\mathcal{W}$ by ℓ.

Lemma 113. *A line refinement of a polyation $\mathcal{W}$ of a polygonal region has the same area sum as $\mathcal{W}$.*

Theorem 114. *All polyations of a polygonal region* $\mathbf{F}$ *have the same area sum. In particular all triangulations of* $\mathbf{F}$ *have that same area sum.*

(Let $\mathcal{P}$ and $\mathcal{Q}$ be any two polyations of $\mathbf{F}$. Find a finite list: $\ell_0, \ell_1, \ldots, \ell_{m-1}$ of lines and sequences $\mathcal{U}_0, \mathcal{U}_1, \mathcal{U}_2, \ldots$ and $\mathcal{V}_0, \mathcal{V}_1, \mathcal{V}_2, \ldots$ of new polyations of $\mathbf{F}$ with the following properties.

1. $\mathcal{U}_0 = \mathcal{P}$ and $\mathcal{V}_0 = \mathcal{Q}$.
2. For $i = 0, 1, \ldots, m-1$ we have $\mathcal{U}_{i+1} = \mathcal{U}_i[\ell_i]$ and $\mathcal{V}_{i+1} = \mathcal{V}_i[\ell_i]$.
3. $\mathcal{U}_m = \mathcal{V}_m$.

It will follow that $\Sigma\mathcal{P} = \Sigma\mathcal{Q}$.)

Using Theorem 114 we can now unambiguously *define* the area measure function $\mathcal{A}$.

For each polygonal region **F** *we define* $\mathcal{A}(\mathbf{F})$, *the* **area** *of* **F**, *to be the area sum of every triangulation of* **F**.

Note that this definition is consistent with our previous definitions of area for triangular regions, convex quadrilateral regions and convex polygonal regions.

4.4 The Area Measure Theorem

Having defined the function $\mathcal{A}$, will will now prove the Area Measure Theorem 120 saying that $\mathcal{A}$ has the basic properties we would expect of any area measure function. These are parts (i), (ii) and (iii) of the Area Measure Axiom of [2]. Section 7.1 will be primarily devoted to a proof of Theorem 199, which is part (iv) of that axiom.

Lemma 115. *The area of a unit square is one.*

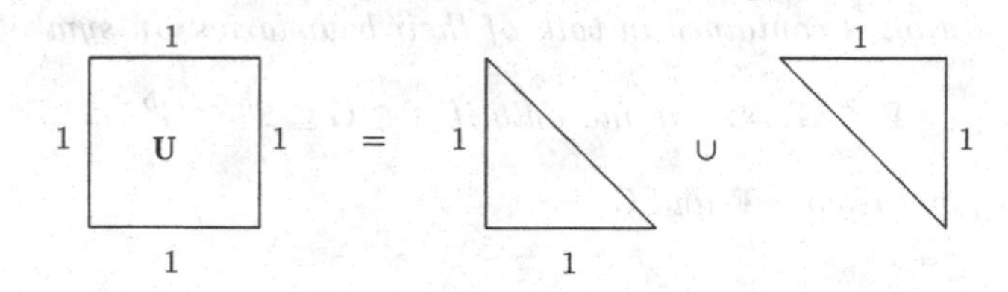

Fig. 4.14 Area of a unit square.

Next we would like to prove that isometries preserve area. Isometries were defined to be bijections that preserve distance. Since the distance function d is our only primitive construct, every other construct has been defined in terms of d and previous constructs. As a result it is a totally routine matter to verify that each of these other constructs is also preserved in an appropriate sense by isometries. The following theorem articulates exactly what this means for a number of key constructs.

Theorem 116. *Isometries preserve length, betweenness, segments, rays, lines, sides of lines, angles, bisectors of angles, right angles and circles, More specifically, if* α *is an isometry,* A, B, C *are distinct points, then*

(i) $\mathcal{L}(AB) = \mathcal{L}(A^\alpha B^\alpha)$,
(ii) $[ABC]$ *if and only if* $[A^\alpha B^\alpha C^\alpha]$,
(iii) $(AB)^\alpha = A^\alpha B^\alpha$,
(iv) $(\overrightarrow{AB})^\alpha = \overrightarrow{A^\alpha B^\alpha}$,
(v) $(\overleftrightarrow{AB})^\alpha = \overleftrightarrow{A^\alpha B^\alpha}$,
(vi) α *of the* C*-side of* $\overleftrightarrow{AB}$ *is the* C^α*-side of* $(\overleftrightarrow{AB})^\alpha$ *if* $C \notin \overleftrightarrow{AB}$.

(vii) $(\angle AOB)^\alpha = \angle A^\alpha O^\alpha B^\alpha$
(viii) *If* $\overrightarrow{OX}$ *bisects* $\angle AOB$, *then* $\overrightarrow{O^\alpha X^\alpha}$ *bisects* $\angle A^\alpha O^\alpha B^\alpha$.
(ix) *If* $\angle AOB$ *is a right angle, then* $\angle A^\alpha O^\alpha B^\alpha$ *is a right angle.*
(x) *If* $\mathbf{C}$ *is a circle with center* O *and radius* r, *then* $\mathbf{C}^\alpha$ *is the circle with center* O^α *and radius* r.

Lemma 117. *Congruent polygonal regions have the same area.*

(Let **F** and **G** be polygonal regions and let φ be an isometry that takes **F** onto **G**. Show that φ maps a set of triangular regions with disjoint interiors whose union is **F** onto a set of triangular regions with disjoint interiors whose union is **G**. Further, show that those two sets of triangular regions have the same area sum.)

The caveat in Theorem 120(iii), that **F** and **G** overlap only along their boundaries, immediately implies that their interiors are disjoint. In fact the converse is true as well.

Lemma 118. *Two polygonal regions have disjoint interiors if and only if their intersection is contained in both of their boundaries. In symbols,*

$$\mathbf{F}^i \cap \mathbf{G}^i = \varnothing \quad \text{if and only if} \quad \mathbf{F} \cap \mathbf{G} \subseteq \mathbf{F}^b \cap \mathbf{G}^b$$

for all polygonal regions **F** *and* **G**.

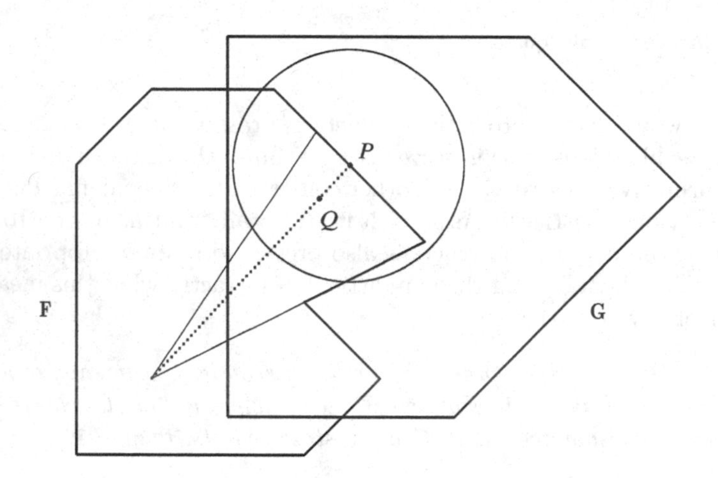

Fig. 4.15 $\mathbf{F}^b \cap \mathbf{G}^i \neq \varnothing$ implies $\mathbf{F}^i \cap \mathbf{G}^i \neq \varnothing$.

(Assume $\mathbf{F} \cap \mathbf{G} \not\subseteq \mathbf{F}^b \cap \mathbf{G}^b$. You need to show that $\mathbf{F}^i \cap \mathbf{G}^i \neq \varnothing$. Note that the intersection $\mathbf{F} \cap \mathbf{G}$ is a disjoint union of four sets:

$$\begin{aligned}\mathbf{F} \cap \mathbf{G} &= (\mathbf{F}^b \cup \mathbf{F}^i) \cap (\mathbf{G}^b \cup \mathbf{G}^i) \\ &= (\mathbf{F}^b \cap \mathbf{G}^b) \cup (\mathbf{F}^b \cap \mathbf{G}^i) \cup (\mathbf{F}^i \cap \mathbf{G}^b) \cup (\mathbf{F}^i \cap \mathbf{G}^i).\end{aligned}$$

Since $\mathbf{F} \cap \mathbf{G} \not\subseteq \mathbf{F}^b \cap \mathbf{G}^b$, it remains to show that, if either $\mathbf{F}^b \cap \mathbf{G}^i \neq \varnothing$ or $\mathbf{F}^i \cap \mathbf{G}^b \neq \varnothing$, then $\mathbf{F}^i \cap \mathbf{G}^i \neq \varnothing$. Assume $\mathbf{F}^b \cap \mathbf{G}^i \neq \varnothing$ contains the point P. (Figure 4.15). Using these facts, locate a point Q in $\mathbf{F}^b \cap \mathbf{G}^i$.)

In practice we find that it is often easier to exhibit a line separating the interiors of **F** and **G** than to prove that their intersection is contained in their boundaries.

Lemma 119. *If the polygonal regions* **F** *and* **G** *have disjoint interiors, then* $\mathbf{F} \cup \mathbf{G}$ *is a polygonal region and*

$$\mathcal{A}(\mathbf{F} \cup \mathbf{G}) = \mathcal{A}(\mathbf{F}) + \mathcal{A}(\mathbf{G}).$$

We can now gather these lemmas into our final theorem.

Area Measure Theorem 120 ([2] Axiom 6). *The area measure function* $\mathcal{A}$ *has the following properties.*

(i) *The area of the unit square is* 1.
(ii) *Congruent polygonal regions have the same area.*
(iii) *If the polygonal regions* **F** *and* **G** *have disjoint interiors, then* $\mathbf{F} \cup \mathbf{G}$ *is a polygonal region and* $\mathcal{A}(\mathbf{F} \cup \mathbf{G}) = \mathcal{A}(\mathbf{F}) + \mathcal{A}(\mathbf{G})$.

4.5 Consequences of Area Measure

We end this chapter by using our definition of area to prove some of the standard area formulas from high school. If b and b' are the lengths of two parallel sides of a trapezoid **T** and h is the distance between them given by Theorem 59, then we call b and b' **bases** for **T** and h the corresponding **height** of **T**.

Theorem 121 ([2] T74). *If* b *and* b' *are bases for a trapezoid* **T** *and* h *is the corresponding height, then*

$$\mathcal{A}(\mathbf{T}) = \tfrac{(b+b')h}{2}.$$

If **P** is a parallelogram, b is the length of a side of **P** and h is the distance between that side and the side parallel to it, then we call b a **base** of **P** and h the corresponding **height** of **P**.

Theorem 122 ([2] T73). *If* b *is a base of a parallelogram* **P** *and* h *is the corresponding height, then*

$$\mathcal{A}(\mathbf{P}) = bh.$$

Rectangle Area Theorem 123 ([2] p32). *If b is a base of a rectangle* **R** *and h is the corresponding height, then*

$$\mathcal{A}(\mathbf{R}) = bh.$$

Theorem 124 ([2] T75).

(i) *The two diagonals of a rhombus* **R** *divide* $\overline{\mathbf{R}}$ *into four congruent right triangular regions with disjoint interiors.*
(ii) *The area of a rhombus* **R** *with diagonals of length d_1 and d_2, is*

$$\mathcal{A}(\mathbf{R}) = \frac{d_1 d_2}{2}.$$

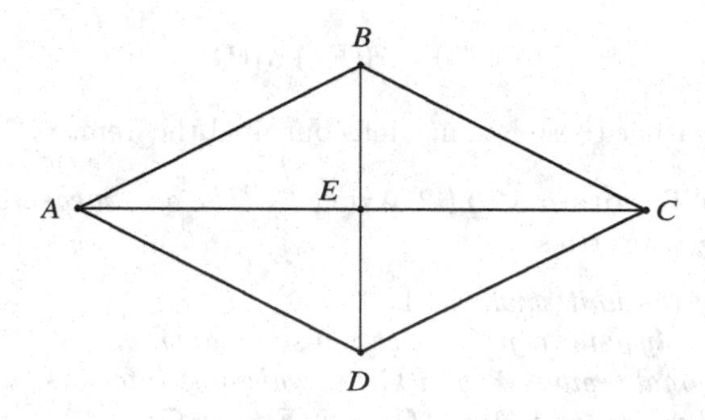

Fig. 4.16 Area of a rhombus. [2]

As a consequence of the Area Measure Theorem 120 we see that, for two polygonal regions, decompositional equivalence is indeed a special case of equal area just as congruence is a special case of decompositional equivalence.

Theorem 125. *Two polygonal regions that are decompositionally equivalent have the same area.*

You can now obtain the full Pythagorean Theorem 92 as an immediate consequence of Pythagorean Theorem 101, the Rectangle Area Theorem 123 and Theorem 125.

4th Pythagorean Theorem 126. *If a right triangle has legs of lengths a and b and hypotenuse of length c, then $c^2 = a^2 + b^2$.*

Chapter 5
Angle Measure

Like area measure, angle measure is a construct that can either be primitive or defined. Elementary treatments, as given in [2], take the degree measure function $\mathcal{D}$ as a primitive construct governed by an axiom asserting that it has properties similar to those of length and area measure. This option provides an efficient shortcut to proving further properties of angle measure and their consequences. But it obligates us to define a measure function $\mathcal{D}$ and prove that the axiom holds for $\mathcal{D}$ in each model before we can conclude that the theorems are true in that model.

Our treatment of angle measure contrasts with the three widely used axiomatic developments of plane Euclidean geometry given in [7], [10] and [11]. All three use the Ruler Postulate and treat hyperbolic geometry as well. All three treat length and polygonal area much as we do. But they do less with angle and circle measure.

- Hartshorne [7] takes distance as the only primitive construct, making only minor references to angle measure, and shows that the real coordinate plane $\mathbb{R}^2$ is a model of his axioms.
- Millman and Parker [10] take distance, lines and angle measure as primitive constructs and show that $\mathbb{R}^2$ is the only model of their axioms. As we describe in Chapter 8, they use differences and inner products of points in $\mathbb{R}^2$ together with the inverse cosine function to define angle measure in $\mathbb{R}^2$.
- Moise [11] takes distance and angle measure as primitive constructs governed by axioms. He then presents $\mathbb{R}^2$ as a candidate for a model, but refrains from defining angle measure in $\mathbb{R}^2$ with the claim (page 439) that doing so is a "formidable technical chore". This makes it difficult to know which of his theorems do not depend on the angle measure axioms on pages 95 and 96 and are therefore true in $\mathbb{R}^2$.

Our title set a clear goal that required us to include a full axiomatic treatment of angle measure. It would have been an option to follow Millman and Parker, giving degree measure as a primitive construct governed by our

D. M. Clark, S. Pathania, *A Full Axiomatic Development of High School Geometry*,
https://doi.org/10.1007/978-3-031-23525-2_5

Angle Measure Theorem 164 stated as an axiom. We could have then shown in Chapter 8 that angle measure could be defined in each model as they do. However, our intention was also to develop geometry from distance measure alone, defining all other constructs by drawing on concepts familiar to high school teachers and their students without making an exception for angle measure.

Our definition of the degree measure $\mathcal{D}$ in Section 5.2 draws on the method of [2] that allows students to approximate the measure of any given angle $\angle BOA$. This method is illustrated in Figure 5.1. It begins by constructing a perpendicular ray from O on the B-side of $\overleftrightarrow{OA}$, forming two 90 degree angles. Bisecting these two angles produces four 45 degree angles. Bisecting those produces eight 22.5 degree angles, and so forth. The result is to construct a protractor that they can use to measure $\angle BOA$ by locating $\overrightarrow{OB}$ between two of these binary rays. The more binary rays they produce, the more accurately they can estimate the measure of $\angle BOA$.

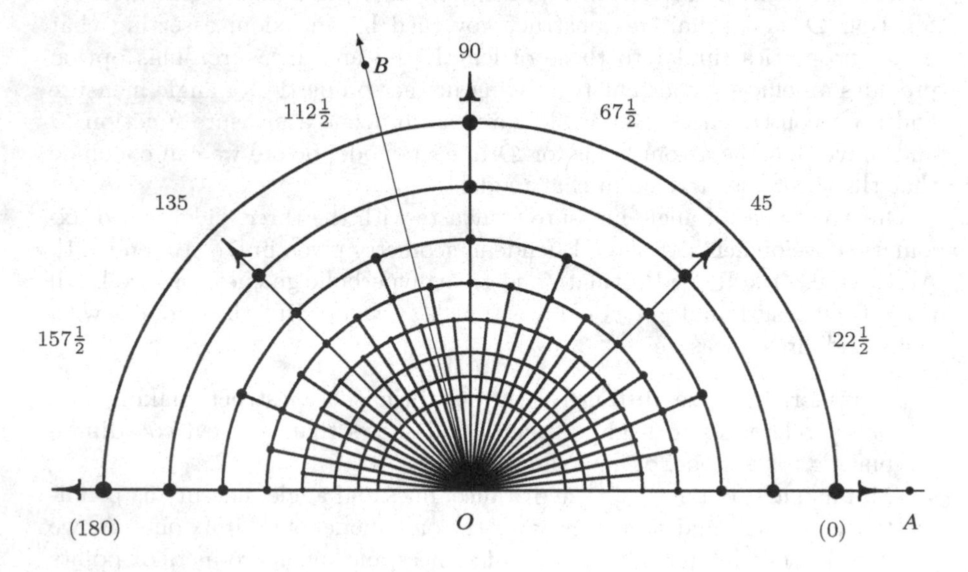

Fig. 5.1 Approximating the measure of $\angle BOA$. [2]

We began our study of area measure with the weaker notion of *same area* that was obtained by decompositional equivalence. Similarly, we will begin our study of angle measure in Section 5.1 with the weaker notion of *angle ordering* that says when one angle is less than another. Our definition of degree measure $\mathcal{D}$ in Section 5.2 is a formalization of the method shown in Figure 5.1. In Sections 5.3 and 5.4 we will prove that $\mathcal{D}$ fulfills the conditions listed in our Angle Measure Theorem 164.

There is just one particularly involved part of the proof of the Angle Measure Theorem 164. Notice that the three parts of that theorem are very close analogs of the first three parts of the Length Measure Axiom (LMA) and to the Area Measure Theorem 120 with one exception. Part (ii) of Theorem 120 says only that two congruent polygonal regions have the same measure. Part (ii) of Theorem 164 says that two angles are congruent if and only if they have the same measure. It turns out that the added assertion in Theorem 164(ii), that

angles with the same measure are congruent,

will take a serious effort to prove. We have therefore allocated Section 5.3 entirely to the proof of Theorem 164(ii).

In Section 5.4 we will complete the proof of the Angle Measure Theorem 164 and we give its standard consequences in Section 5.5. This will include a final proof of the Pythagorean Theorem 175 that uses both area measure and angle measure.

5.1 Ordering Angles

In this section we will apply the method of ordering angles from small to large as a first step toward defining a degree measure on angles in the next section.

Lemma 127. *If $\angle AOC$ is an angle and point B lies on line $\overleftrightarrow{AC}$, then B is interior to $\angle AOC$ if and only if* $[ABC]$.

We will write "$(ABC)_O$" to mean that point B is in the interior of angle $\angle AOC$. Intuitively we can think of the statement "$(ABC)_O$" as saying that ray $\overrightarrow{OB}$ is between the rays $\overrightarrow{OA}$ and $\overrightarrow{OC}$. More generally, if $\angle A_{m-1}OA_0$ is an angle, we will write

$$\text{“}(A_{m-1}A_{m-2}A_{m-3}\dots A_1A_0)_O\text{”}$$

to mean that $(A_kA_jA_i)_O$ for all integers k, j, i such that $0 \le i < j < k < m$. It turns out that the ternary relation $(\dots)_O$ has a property that is a direct analog of the betweenness property 4Pt1.

Lemma 128. *If $(ABC)_O$ and $(ACD)_O$, then $(ABCD)_O$.*

Notice that the analog of the other four point property 4Pt2 is clearly false: $(ABC)_O$ and $(BCD)_O$ do *not* imply $(ABCD)_O$.

We would like to say that $\angle A'O'B'$ is **less than** $\angle AOB$, written

$$\angle A'O'B' < \angle AOB$$

(or $\angle AOB > \angle A'O'B'$), if there is a point X such that $(AXB)_O$ and $\angle A'O'B' \cong \angle AOX$. However, this definition is potentially ambiguous since it does not explain which of the two rays $\overrightarrow{OA}$ or $\overrightarrow{OB}$ is to be combined with $\overrightarrow{OX}$ to match $\angle A'O'B'$. It turns out that this is not a problem. The following lemma shows that angle order is indeed well defined by showing that there is an X for which $(AXB)_O$ and $\angle AOX \cong \angle A'O'B'$ if and only if there is a Y for which $(AYB)_O$ and $\angle BOY \cong \angle A'O'B'$.

Lemma 129. *Let X be a point in the interior of $\angle AOB$. Then there is a point Y in the interior of $\angle AOB$ such that $\angle AOX \cong \angle BOY$.*

Theorem 130. *If O is a point, then the relation $(...)_O$ is preserved by isometries. In symbols, let A, O, B be non-collinear points, let X be a point and let α be an isometry. Then*

$$(AXB)_O \text{ if and only if } (A^\alpha X^\alpha B^\alpha)_{O^\alpha}.$$

Lemma 131. *If $\angle A$, $\angle B$ and $\angle C$ are angles, then*

(i) $\angle A \not< \angle A$,
(ii) $\angle A < \angle B < \angle C$ *implies* $\angle A < \angle C$,
(iii) $\angle A < \angle B \cong \angle C$ *implies* $\angle A < \angle C$
$\angle A > \angle B \cong \angle C$ *implies* $\angle A > \angle C$ *and*
(iv) *exactly one of* $\angle A < \angle B$ *or* $\angle A \cong \angle B$ *or* $\angle A > \angle B$ *is true.*

Theorem 132. *Isometries preserve angle order. In symbols, let $\angle A$ and $\angle B$ be angles and let and α be an isometry. Then*

$$\angle A < \angle B \text{ if and only if } (\angle A)^\alpha < (\angle B)^\alpha.$$

We can now give a refinement of Theorem 40, which said that no angle of a triangle is congruent to an opposite exterior angle.

Theorem 133. *Each angle of a triangle is less than each opposite exterior angle.*

Lemma 134. *Let A, B, C be three collinear points. If $B \in \overrightarrow{AC}$ and $\mathcal{L}(AB) < \mathcal{L}(AC)$, then $[ABC]$.*

Theorem 135. *In a triangle, the larger side is opposite the larger angle and the larger angle is opposite the larger side.*

Theorem 136. *Let A and B be two points on a circle* **C** *with center O and radius r, and let X be a third point of line $\overleftrightarrow{AB}$. Then X is inside* **C** *if it is between A and B and is outside* **C** *otherwise.*

5.2 Degree Measure

Building on the facts we have established about ordering of angles, we will describe a way to assign a number between 0 and 180 to each angle that we will call its *degree measure*. To do this we begin by fixing an arbitrary angle $\angle BOA$ and using binary rays close to $\overrightarrow{OB}$ to define its degree measure.

Lemma 137. *Let P_x, P_y and P_z be points of a circle* **C** *with center O such that P_x and P_z are not collinear with O. If $(P_xP_yP_z)_O$, then neither P_x and P_y nor P_z and P_y are collinear with O.*

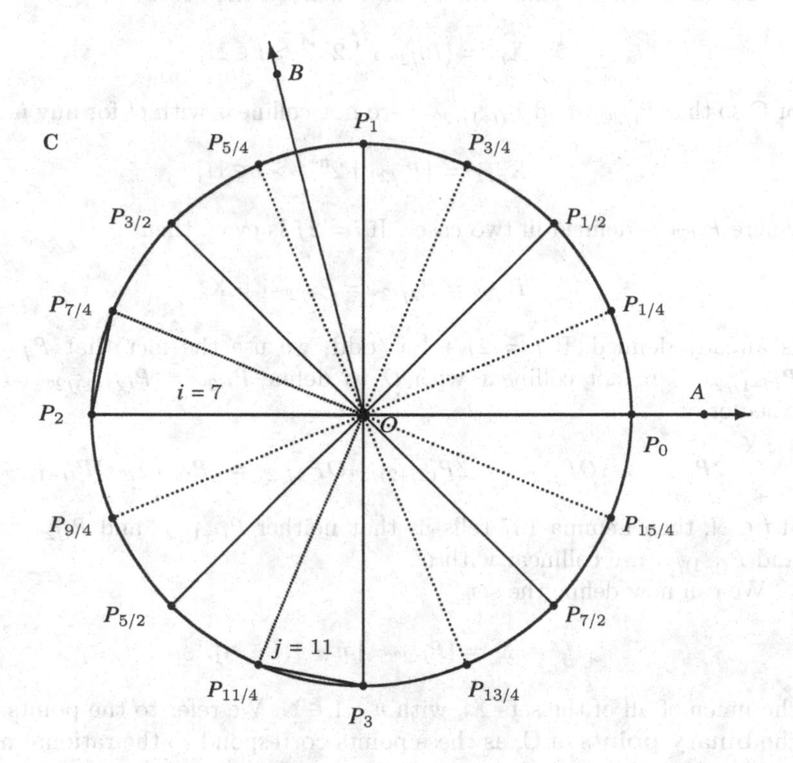

Fig. 5.2 Binary points $\mathbf{X}_3$ from $\mathbf{X}_2$.

Now consider the fixed angle $\angle BOA$ and let **C** be any fixed circle with center O. We use the letter "$\mathbb{N}$" to denote the *natural numbers*:

$$\mathbb{N} := \{0, 1, 2, 3, \dots\}.$$

We will recursively define a nested sequence

$$\mathbf{X}_1 \subseteq \mathbf{X}_2 \subseteq \mathbf{X}_3 \subseteq \mathbf{X}_4 \dots$$

of subsets of $\mathbf{C}$ as follows. Let P_0 be the intersection of $\overrightarrow{OA}$ with $\mathbf{C}$ and let P_1 be the point of $\mathbf{C}$ on the B-side of $\overleftrightarrow{OA}$ such that $\overleftrightarrow{OP_1}$ is perpendicular to $\overleftrightarrow{OP_0}$. (See Figure 5.2.) Choose P_2 and P_3 on $\mathbf{C}$ so that $[P_2OP_0]$ and $[P_1OP_3]$. From here onward, if x is a number and we define a point P_x, we will understand "P_{x+4}" to be an alternate name for P_x. We now define

$$\mathbf{X}_1 := \{P_0, P_1, P_2, P_3\} = \{P_{i/2^{1-1}} \mid 2^{1+1} > i \in \mathbb{N}\}.$$

Note that if $i \in \mathbb{N}$, then $\angle P_{i+1}OP_i$ is a right angle so that P_i and P_{i+1} are not collinear with O.

Assume $0 < n \in \mathbb{N}$ and that we have defined the subset

$$\mathbf{X}_n := \{P_{i/2^{n-1}} \mid 2^{n+1} > i \in \mathbb{N}\}$$

of $\mathbf{C}$ so that $P_{i/2^{n-1}}$ and $P_{(i+1)/2^{n-1}}$ are not collinear with O for any $i \in \mathbb{N}$. Let

$$\mathbf{X}_{n+1} := \{P_{i/2^n} \mid 2^{n+2} > i \in \mathbb{N}\}$$

where $P_{i/2^n}$ is defined in two cases. If $i = 2j$ is even, then

$$P_{i/2^n} = P_{2j/2^n} = P_{j/2^{n-1}} \in \mathbf{X}_n$$

is already defined. If $i = 2j + 1$ is odd, we use the fact that $P_{j/2^{n-1}}$ and $P_{(j+1)/2^{n-1}}$ are not collinear with O to define $P_{i/2^n} = P_{(2j+1)/2^n}$ to be the bisector of

$$\angle P_{(j+1)/2^{n-1}}OP_{j/2^{n-1}} = \angle P_{(2j+2)/2^n}OP_{2j/2^n} = \angle P_{(i+1)/2^n}OP_{(i-1)/2^n}.$$

If $i \in \mathbb{N}$, then Lemma 137 tells us that neither $P_{(i+1)/2^n}$ and $P_{i/2^n}$ nor $P_{i/2^n}$ and $P_{(i-1)/2^n}$ are collinear with O.

We can now define the set

$$\mathbf{X} := \{P_{i/2^{n-1}} \mid n-1, i \in \mathbb{N}\},$$

the union of all of the sets $\mathbf{X}_n$ with $n - 1 \in \mathbb{N}$. We refer to the points of $\mathbf{X}$ as the **binary points** of $\mathbf{C}$, as these points correspond to the rational numbers $x \in [0, 4)$ which have a finite decimal representation when written base two.

Lemma 138 ([2] T83). *Assume that $\angle ABC \cong \angle A'B'C'$. Let $\overrightarrow{BD}$ bisect $\angle ABC$ and $\overrightarrow{B'D'}$ bisect $\angle A'B'C'$. Then $\angle ABD \cong \angle A'B'D'$.*

Theorem 139. *Given the angle $\angle BOA$, the circle $\mathbf{C}$ with center O and its associated binary points $\mathbf{X}$ defined above, we have, for all i, j and n with $n - 1, i, j \in \mathbb{N}$,*

$$\triangle P_{i/2^{n-1}}OP_{(i+1)/2^{n-1}} \cong \triangle P_{j/2^{n-1}}OP_{(j+1)/2^{n-1}}.$$

In particular,

$$\angle P_{i/2^{n-1}}OP_{(i+1)/2^{n-1}} \cong \angle P_{j/2^{n-1}}OP_{(j+1)/2^{n-1}}$$

and

$$\mathcal{L}(P_{i/2^{n-1}}P_{(i+1)/2^{n-1}}) = \mathcal{L}(P_{j/2^{n-1}}P_{(j+1)/2^{n-1}}).$$

Looking at Figure 5.2 we would like to define degree measure in such a way that the measure of $\angle P_1OP_0$ is 90, the measure of $\angle P_{3/4}OP_0$ is $90 \times \frac{3}{4} = 67\frac{1}{2}$ and the measure of $\angle P_{i/2^{n-1}}OP_0$ is $90i/2^{n-1}$ for every binary point $P_{i/2^{n-1}}$ with $i/2^{n-1} < 2$. More generally, we would like to define the degree measure of the arbitrary angle $\angle BOA = \angle BOP_0$ in such a way that, for $0 < i < 2^n$, the degree measure of $\angle BOA$ is more than $90i/2^{n-1}$ if and only if $P_{i/2^{n-1}}$ is in the interior of $\angle BOA$. Figure 5.2 suggests that we define the *degree measure* of $\angle BOA$ as 90 times the least upper bound of the set of $i/2^{n-1}$ such that $i < 2^n$ and $P_{i/2^{n-1}}$ is in the interior of $\angle BOA$; in symbols,

$$\mathcal{D}(\angle BOA) := 90\,\text{lub}\left\{\tfrac{i}{2^{n-1}} \,\middle|\, n-1 \in \mathbb{N}, i < 2^n \text{ and } (BP_{i/2^{n-1}}A)_O\right\}.$$

For example, Figure 5.1 suggests that the degree measure of $\angle BOA$ is somewhere between $90 \times \frac{18}{16} = 101\frac{1}{4}$ and $90 \times \frac{19}{16} = 106\frac{7}{8}$.

As was the case with the definition of ordering of angles, we need to establish that this definition does not depend on the choices we must make to apply it. To measure $\angle BOA$ there are two choices we need to make. First we need to choose a circle $\mathbf{C}$ with center O. Secondly, we need to choose whether to take P_0 on $\overrightarrow{OA}$, thereby measuring $\angle BOA$, or to take P_0 on $\overrightarrow{OB}$, thereby measuring $\angle AOB$. We need to establish that neither of these choices will change the value of the degree measure.

Lemma 140. *Assume $\mathbf{C}$ and $\mathbf{C}'$ are two circles centered at O. Take P_0 and P_1 to be points on $\mathbf{C}$ such that $\angle P_1OP_0$ is a right angle and let $\mathbf{X} = \{P_{i/2^{n-1}} \mid i, n-1 \in \mathbb{N}\} \subseteq \mathbf{C}$ be the resulting set of binary points. For $k = 0, 1$ let P'_k be the intersection of $\overrightarrow{OP_k}$ with $\mathbf{C}'$ and let $\mathbf{X}' = \{P'_{i/2^{n-1}} \mid i, n-1 \in \mathbb{N}\} \subseteq \mathbf{C}'$ be the set of binary points obtained by starting with P'_0 and P'_1. Then, for each positive integer n, we have*

$$\overrightarrow{OP_{i/2^{n-1}}} = \overrightarrow{OP'_{i/2^{n-1}}}$$

for all $i \in \mathbb{N}$.

Lemma 141. *The value of $\mathcal{D}(\angle BOA)$ is independent of the choice of circle $\mathbf{C}$ centered at O that is used to calculate it.*

Lemma 142. *Let $\mathbf{C}$ be a circle with center O and binary points*

$$\mathbf{X} = \{P_{i/2^{n-1}} \mid i, n-1 \in \mathbb{N}\}$$

and let α be an isometry. Let

$$\mathbf{X}' = \{P'_{i/2^{n-1}} \mid i, n-1 \in \mathbb{N}\}$$

be the set of binary points for the circle $\mathbf{C}^\alpha$ *with center* O^α *obtained by starting with points* $P'_0 := P^\alpha_0$ *and* $P'_1 := P^\alpha_1$ *of* $\mathbf{C}^\alpha$ *forming the right angle* $\angle P'_1 O^\alpha P'_0$. *Then, for all* $j, n-1 \in \mathbb{N}$, *we have*

$$P'_{j/2^{n-1}} = P^\alpha_{j/2^{n-1}}.$$

Lemma 143. *Let* $\angle O$ *be an angle that is the union of two rays* $\overrightarrow{OA}$ *and* $\overrightarrow{OB}$. *Then the value of the degree measure of angle* $\angle O$ *is the same whether we calculate it using*

(i) P_0 *on* $\overrightarrow{OA}$ *and* P_1 *on the* B*-side of* $\overleftrightarrow{OA}$ *to obtain* $\mathcal{D}(\angle BOA)$ *or*
(ii) P_0 *on* $\overrightarrow{OB}$ *and* P_1 *on the* A*-side of* $\overleftrightarrow{OB}$ *to obtain* $\mathcal{D}(\angle AOB)$.

Theorem 144. *Let* $\mathbf{X}$ *be a set of binary points with* $A = P_0$ *and* B *on the* P_1*-side of* $\overleftrightarrow{OA}$. *Then the* **degree measure** *of* $\angle BOA$ *is well defined as*

$$\mathcal{D}(\angle BOA) := 90\,\mathrm{lub}\left\{\tfrac{i}{2^{n-1}} \,\middle|\, n-1 \in \mathbb{N}, i < 2^n \text{ and } (BP_{i/2^{n-1}}A)_O\right\}.$$

The number 90 in the definition of $\mathcal{D}$ has historical roots but has no mathematical significance. For any positive number u, we could define u-measure $\mathcal{M}_u$ by replacing 90 with u in the definition of degree measure. This would give us

$$\mathcal{M}_u(\angle AOB) = \tfrac{u}{90}\mathcal{D}(\angle AOB)$$

for every angle $\angle AOB$. It would then be easy to check that the restatement of each of the degree measure theorems we obtain, with $\mathcal{M}_u$ replacing $\mathcal{D}$ and u replacing 90, would follow as immediate corollaries of those theorems. For example, taking $u = \pi/2$ would give us radian measure $\mathcal{R}$ and the familiar conversion

$$\mathcal{R}(\angle AOB) = \tfrac{\pi}{180}\mathcal{D}(\angle AOB).$$

5.3 Angle Measure and Congruence

This section is devoted to the final goal of proving Theorem 158 stating that two angles are congruent if and only if they have the same degree measure. One direction is a fairly direct consequence of what we already know from Theorem 116 to be preserved by isometries.

Theorem 145. *Isometries preserve degree measure. In symbols, if* α *is an isometry and* $\angle AOB$ *is an angle, then*

$$\mathcal{D}(\angle AOB) = \mathcal{D}(\angle A^\alpha O^\alpha B^\alpha).$$

It remains to prove the converse, that angles with the same degree measure are congruent. Accordingly we consider a pair of angles with the same degree measure:

$$\mathcal{D}(\angle AOB) = \mathcal{D}(\angle A'O'B').$$

In order to use the fact that these measures are the same, we need to lay out the context that says how the angles are measured. Let r be any fixed number in $\mathbb{F}^+$ and consider the circles $\mathbf{C} := \mathbf{C}_r^O$ and $\mathbf{C}' := \mathbf{C}_r^{O'}$. By Corollary 8 we can assume that $A, B \in \mathbf{C}$ and $A', B' \in \mathbf{C}'$. Let $\mathbf{X}$ and $\mathbf{X}'$ be systems of binary points with

$$A = P_0 \in \mathbf{X} \subseteq \mathbf{C} \text{ and } A' = P_0' \in \mathbf{X}' \subseteq \mathbf{C}'$$

and with P_1 on the B-side of $\overleftrightarrow{OA}$ and P_1' on the B'-side of $\overleftrightarrow{O'A'}$, as illustrated in Figure 5.3.

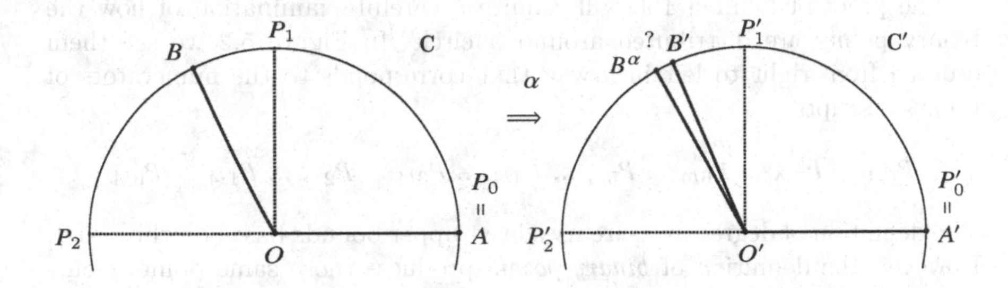

Fig. 5.3 $\mathcal{D}(\angle BOA) = \mathcal{D}(\angle B'O'A')$.

The task before us is to find an isometry β such that

$$A^\beta = A', \quad O^\beta = O', \quad \text{and} \quad B^\beta = B'.$$

Then we would have $\angle AOB \cong \angle A'O'B'$ by Theorem 116. The following lemma comes close to giving us a successful choice for β.

Lemma 146. *Given the configuration of Figure 5.3, there is an isometry α such that*

$$A^\alpha = A', \quad O^\alpha = O' \quad \textit{and} \quad B^\alpha \textit{ is on the } B'\textit{-side of } \overleftrightarrow{O'A'}.$$

Since this isometry α preserves degree measure (Theorem 145), we have

$$\mathcal{D}(\angle A'O'B') = \mathcal{D}(\angle AOB) = \mathcal{D}(A'O'B^\alpha).$$

Since we also have B^α on the B'-side of $\overleftrightarrow{O'A'}$, it would appear from Figure 5.3 that B^α should be B'. If it is, then $\beta := \alpha$ is the isometry we are seeking.

However, our only knowledge of B' is that it produces an angle at O' of the same degree measure as B produces at O. Consequently any proof that $B^\alpha = B'$ will need to use the definition of degree measure. Suppose, for example, that B' and B^α were two different points that were close enough so that there were no binary points in the interior of $\angle B'O'B^\alpha$. Then the definition of degree measure would say that

$$\mathcal{D}(\angle A'O'B') = \mathcal{D}(\angle A'O'B^\alpha),$$

but Theorem 35 would tell us that

$$\angle(A'O'B') \not\cong \angle(A'O'B^\alpha).$$

This shows that any proof of Theorem 158 will necessarily imply that every central angle of $\mathbf{C}'$ must have binary points in its interior. Conversely, you can check now that this fact is our Lemma 156 and that Theorem 158 immediately follows.

The proof of Lemma 156 will require a careful examination of how the binary points are distributed around a circle. In Figure 5.2 we see them ordered from right to left in a way that corresponds to the numerators of their subscripts:

$$P_{8/4} \;\dots\; P_{7/4} \;\dots\; P_{6/4} \;\dots\; P_{5/4} \;\dots\; P_{4/4} \;\dots\; P_{3/4} \;\dots\; P_{2/4} \;\dots\; P_{1/4} \;\dots\; P_{0/4}$$

The definition of degree measure uses least upper bounds based on this order. However, the definition of *binary points* produces those same points recursively in a very different order that is based on the denominators of their subscripts:

$$P_{0/1}, P_{1/1}, P_{2/1}, P_{1/2}, P_{3/2}, P_{1/4}, P_{3/4}, P_{5/4}, P_{7/4}, P_{1/8}, P_{3/8}, P_{5/8}, P_{7/8} \dots .$$

Our proofs of facts about the measures of angles formed by binary points will need to use induction on their denominators, this second order. Yet they will also need to use the definition of degree measure which is based on their numerators, the first order. Bridging the gap between these two orders will be the central theme of this section.

Question. *Let $0 < i < j < 2^n$ with $i, j, n \in \mathbb{N}$. Can you guess the degree measure of each of the following angles?*

$$\angle P_1 O P_0, \quad \angle P_{1/2} O P_0, \quad \angle P_{(i+1)/2^{n-1}} O P_{i/2_{n-1}}, \quad \angle P_{j/2^{n-1}} O P_{i/2^{n-1}}.$$

Proving Lemma 156 will require finding proofs of the answers to these questions. That process begins with a lemma about the ordering of binary points with the same denominator.

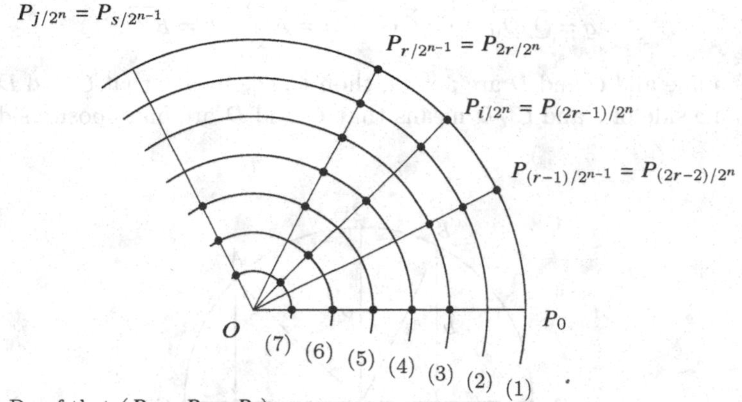

Fig. 5.4 Proof that $(P_{j/2^n}P_{i/2^n}P_0)_O$.

Lemma 147. *Let* **C** *be a circle with center* O *and let* **X** *be a set of binary points on* **C**. *If* $n-1 \in \mathbb{N}$ *and* $0 < i, j < 2^n$, *then*

$$(P_{j/2^{n-1}}P_{i/2^{n-1}}P_0)_O \quad \textit{if and only if} \quad i < j.$$

(Prove this by induction on n. Assuming it is true for $n-1$, prove it for n by considering different possible cases. In each case you will want to use Lemma 128 and the definition of binary points. We give here the proof for one of those cases in seven steps that are illustrated in Figure 5.4. This figure shows five rays for the five relevant points and seven arcs for the seven steps. Three points X, Y, Z on an arc indicate that $(X, Y, Z)_O$.

Case. $i = 2r-1$ is odd, $j = 2s$ is even and $r < s$. Then $1 < r < s < 2^{n-1}$ and we reason as follows.

1. $(P_{2r/2^n}P_{(2r-1)/2^n}P_{(2r-2)/2^n})_O$ since $P_{(2r-1)/2^n}$ is on the bisector.
2. $(P_{r/2^{n-1}}P_{i/2^n}P_{(r-1)/2^{n-1}})_O$ by (1).
3. $(P_{r/2^{n-1}}P_{(r-1)/2^{n-1}}P_0)_O$ by induction since $r-1 > 0$.
4. $(P_{r/2^{n-1}}P_{i/2^n}P_0)_O$ by Lemma 128.
5. $(P_{s/2^{n-1}}P_{r/2^{n-1}}P_0)_O$ by induction.
6. $(P_{s/2^{n-1}}P_{i/2^n}P_0)_O$ by Lemma 128.
7. $(P_{j/2^n}P_{i/2^n}P_0)_O$ by (6).

Give a similar argument for each other case.)

The next two lemmas and theorem provide the key to connecting the order of binary points around the circle with the order of their projections onto the segments P_2P_0 and P_3P_1. To address projections to both of those segments we consider a circle **C** with center O, diameter Q_2Q_0 and a point Q_1 on **C** for which $\overleftrightarrow{OQ_1} \perp \overleftrightarrow{Q_2Q_0}$ (Figure 5.5). If D is a point of **C**, we denote by D^* the foot of the perpendicular from D to $\overleftrightarrow{Q_2Q_0}$. We take A and B to be two points of **C** on the Q_1-side of $\overleftrightarrow{Q_2Q_0}$ and give short names to four lines:

$$q = \overleftrightarrow{Q_2Q_0} \quad o = \overleftrightarrow{OQ_1} \quad a = \overleftrightarrow{AA^*} \quad b = \overleftrightarrow{BB^*}.$$

If ℓ is a line and C and D are points, then $C, D|_\ell$ means that C and D are on the same side of ℓ and $C|_\ell D$ means that C and D are on opposite sides of ℓ.

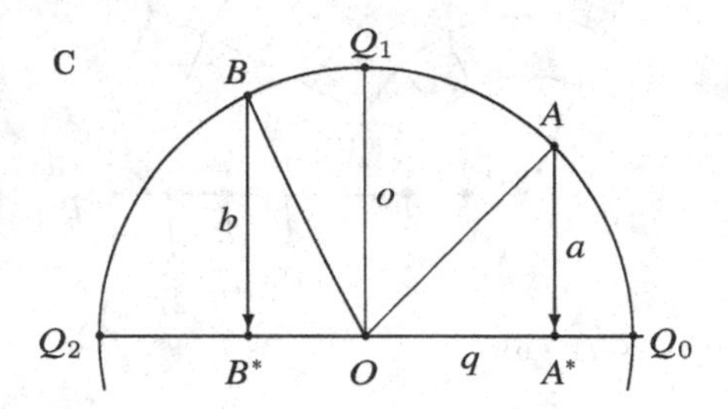

Fig. 5.5 Lines and projections.

Lemma 148. *Let A and B be points on the Q_1-side of q. If $|_o A, Q_0$ and $(BAQ_0)_O$, then $[B^*A^*Q_0]$.*

(These hypotheses and conclusions give a lot of facts about sides of various lines. Pick out from them what you need for the proof.)

Lemma 149. *Let A, B, C be three points of $\mathbf{C}$ on the same side of $\overleftrightarrow{Q_2Q_0}$. Then either $(ABC)_O$ or $(BCA)_O$ or $(CAB)_O$.*

Theorem 150. *If A and B are points of $\mathbf{C}$ such that $(BAQ_0)_O$, then $[B^*A^*Q_0]$.*

(Consider three cases: $A, Q_0|_o$ or $A \in o$ or $A, Q_2|_o$. Use the first to prove the third.)

We can now return to the context in which $\mathbf{C}$ is a circle with center O and $\mathbf{X}_n$ is a set of 2^{n+1} binary points for each positive integer n. Recall from Theorem 139 that the segments formed by successive points of $\mathbf{X}_n$ all have the same length. We will denote that length by

$$s_n := \mathcal{L}(P_{i/2^{n-1}}P_{(i+1)/2^{n-1}}).$$

We would like to show that s_n is small when n is large. To do this it will be sufficient to focus primarily on the first quadrant of $\mathbf{C}$. For each point A on $\mathbf{C}$ let $A^{\downarrow}$ be the foot of the perpendicular from A to $\overleftrightarrow{OP_0}$ and $A^{\leftarrow}$ be the foot of the perpendicular from A to $\overleftrightarrow{OP_1}$. (See Figure 5.6.) For example,

$$P_1^{\downarrow} = O,\ P_0^{\downarrow} = P_0,\ P_1^{\leftarrow} = P_1 \text{ and } P_0^{\leftarrow} = O.$$

In Lemma 147 we saw that the order of binary points of $\mathbf{X}_n$ around the circle matched the order of their subscripts. Combining that fact with

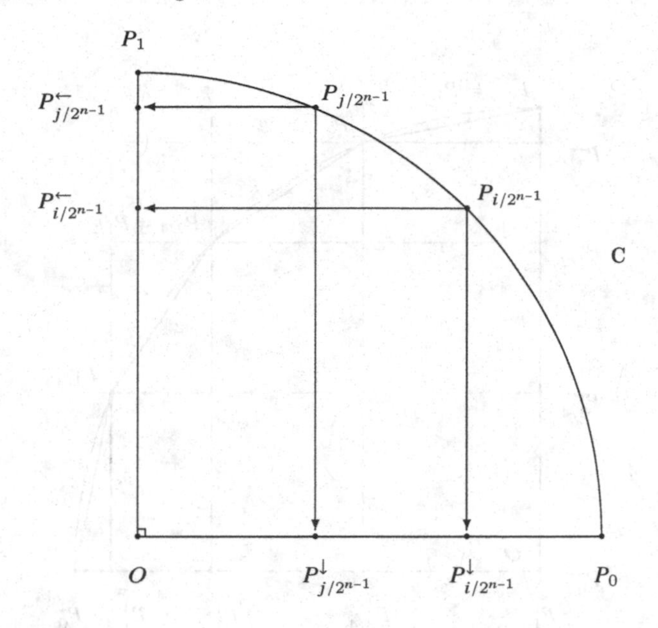

Fig. 5.6 Projections onto OP_0 and OP_1.

Lemma 148 you can now show that the order of their projections on OP_0 and OP_1 also match the order of their subscripts.

Theorem 151. *Let* **C** *be a circle with center* O *and let* **X** *be a set of binary points on* **C**. *If* $n-1 \in \mathbb{N}$ *and* $2^{n-1} > j > i > 0$, *then*

$$\text{(i) } [OP^{\downarrow}_{j/2^{n-1}}P^{\downarrow}_{i/2^{n-1}}P_0] \quad \text{and} \quad \text{(ii) } [OP^{\leftarrow}_{i/2^{n-1}}P^{\leftarrow}_{j/2^{n-1}}P_1].$$

(Use Lemma 147 and Theorem 150 with appropriate choices for Q_0, Q_1 and $*$.)

Theorem 151 tells us that the projections of the binary points from P_0 to O and from O to P_1 lie in exactly the order of their numerical subscripts, justifying the illustration in Figure 5.7 for $n = 3$. This fact allows us to give a tight upper bound for the length s_n of the segments connecting adjacent binary points of $\mathbf{X}_n$.

Theorem 152. *Let* **C** *be a circle with center* O, *radius* r *and binary points* **X** *and let* $2 < n \in \mathbb{N}$ *with* $i < 2^n$. *Then*

$$s_n = \mathcal{L}(P_{i/2^{n-1}}P_{(i+1)/2^{n-1}}) < \tfrac{r}{2^{n-2}}.$$

(Use Theorem 151 and the Triangle Inequality, referring to Figure 5.7.)

Lemma 153. *For every positive integer* n, *each point not on any angle* $\angle P_{(i+1)/2^{n-1}}OP_{i/2^{n-1}}$ *formed by two consecutive points of* $\mathbf{X}_n$ *is in the interior of exactly one of those angles.*

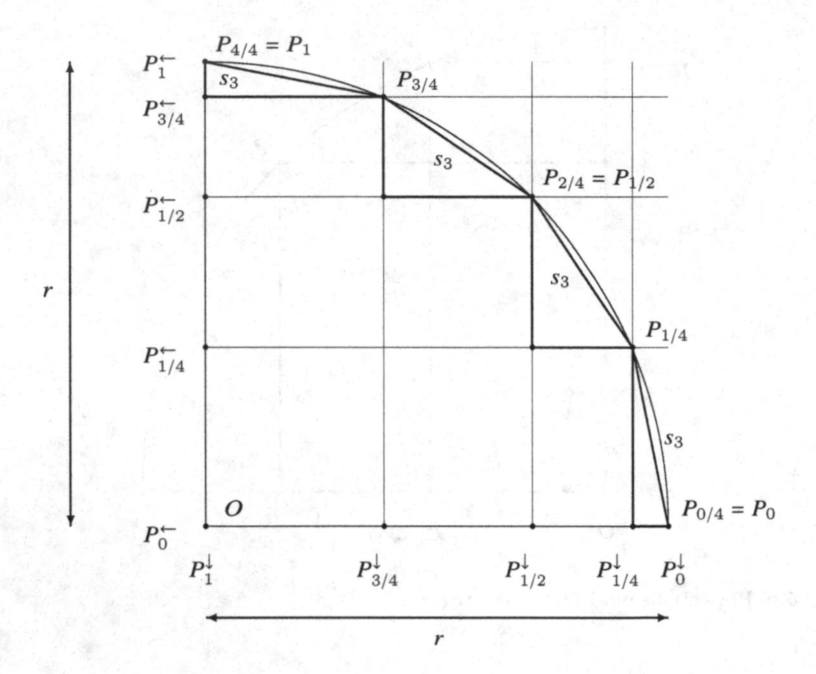

Fig. 5.7 Proof of Theorem 152 for $n = 3$.

(Use induction on n.)

Lemma 154. *Let* **C** *be a circle with center O and let A, B, C be points of* **C** *such that $(ABC)_O$. Then $\mathcal{L}(AB) < \mathcal{L}(AC)$.*

(Use Theorem 135.)

Lemma 155. *If A and B are points of* **C** *on the same side of $\overleftrightarrow{P_2P_0}$ with $(BAP_0)_O$, then there is a binary point in the interior of $\angle BOA$.*

Lemma 156. *If A and B are points of* **C** *on the same side of $\overleftrightarrow{P_2P_0}$ with $(BAP_0)_O$, then there are at least two binary points in the interior of $\angle BOA$.*

This fact has two important consequences. First, we can now show that the ordering of angles can be defined in terms of degree measure.

Theorem 157. *One angle is less than another if and only if it has smaller degree measure. In symbols,*

$$\angle A < \angle B \quad \text{if and only if} \quad \mathcal{D}(\angle A) < \mathcal{D}(\angle B).$$

We can also see that congruence of angles can be defined in terms of degree measure.

Theorem 158. *Angles are congruent if and only if they have the same degree measure. In symbols,*

$$\angle A \cong \angle B \text{ if and only if } \mathcal{D}(\angle A) = \mathcal{D}(\angle B).$$

5.4 The Angle Measure Theorem

We now have the tools we need to prove our Angle Measure Theorem 164. This theorem is the analog of the Length Measure Axiom (LMA) for length measure and the Area Measure Theorem 120 for area measure. As part (ii) of this theorem is our Theorem 158, it remains to prove parts (i) and (iii).

Lemma 159. *If $P_{i/2^{n-1}}$ is a binary point with n a positive integer and $0 < i < 2^n$, then*

$$\mathcal{D}(\angle P_{i/2^{n-1}} O P_0) = \tfrac{90i}{2^{n-1}}.$$

Lemma 160 (Angle Measure (i)). *Every right angle has degree measure 90.*

The proof of Angle Measure (iii) will require two lemmas. The first says that certain rotations are isometries.

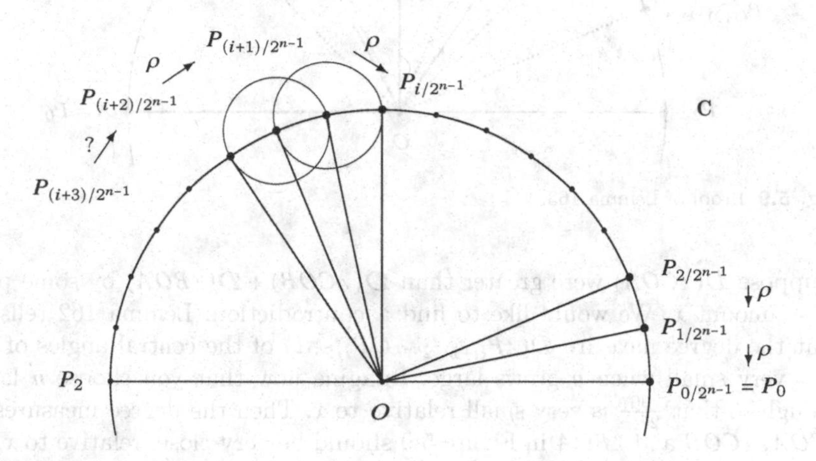

Fig. 5.8 Proof of Lemma 161.

Lemma 161. *Let $0 < n \in \mathbb{N}$. Then there is an isometry ρ such that $O^\rho = O$ and $P^\rho_{(i+1)/2^{n-1}} = P_{i/2^{n-1}}$ for all $i \in \mathbb{N}$.*

(Let ρ be the congruence witnessing the fact that

$$\triangle P_{2/2^{n-1}} O P_{1/2^{n-1}} \cong \triangle P_{1/2^{n-1}} O P_{0/2^{n-1}}.$$

Prove by induction on i that $P_{(i+2)/2^{n-1}} \overset{\rho}{\mapsto} P_{(i+1)/2^{n-1}} \overset{\rho}{\mapsto} P_{i/2^{n-1}}$ for all $i \in \mathbb{N}$. See Figure 5.8.)

Lemma 162. *If* $0 \leq i < j < 2^n$, *then*

$$\mathcal{D}(\angle P_{j/2^{n-1}} O P_{i/2^{n-1}}) = \frac{90(j-i)}{2^{n-1}}$$

(Find an isometry α such that $P_{j/2^{n-1}} \overset{\alpha}{\mapsto} P_{(j-i)/2^{n-1}}$, that $O \overset{\alpha}{\mapsto} O$ and that $P_{i/2^{n-1}} \overset{\alpha}{\mapsto} P_0$.)

Lemma 163 (Angle Measure (iii)). *If B is in the interior of $\angle AOC$, then* $\mathcal{D}(\angle AOC) = \mathcal{D}(\angle AOB) + \mathcal{D}(\angle BOC)$.

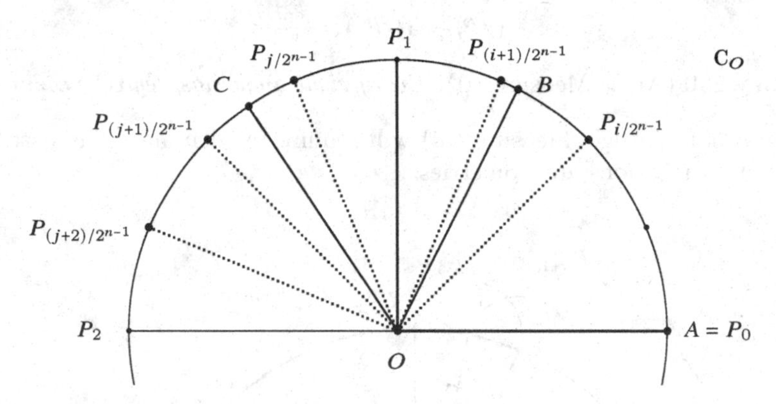

Fig. 5.9 Proof of Lemma 163.

(Suppose $\mathcal{D}(\angle COA)$ were greater than $\mathcal{D}(\angle COB) + \mathcal{D}(\angle BOA)$ by some positive amount x. We would like to find a contradiction. Lemma 162 tells us that the degree measure $\mathcal{D}(\angle P_{(i+1)/2^{n-1}} O P_{i/2^{n-1}})$ of the central angles of $\mathbf{X}_n$ gets very small when n grows large. Imagine now that you choose n large enough so that $\frac{90}{2^{n-1}}$ is very small relative to x. Then the degree measures of $\angle COA$, $\angle COB$ and $\angle BOA$ in Figure 5.9 should be very close, relative to x, to the degree measures of $\angle P_{(j+1)/2^{n-1}} OA$, $\angle P_{j/2^{n-1}} O P_{(i+1)/2^{n-1}}$ and $\angle P_{i/2^{n-1}} OA$, respectively. Using Lemma 162 you can then calculate the difference

$$\mathcal{D}(\angle P_{(j+1)/2^{n-1}} OA) - (\mathcal{D}(\angle P_{j/2^{n-1}} O P_{(i+1)/2^{n-1}}) + \mathcal{D}(\angle P_{i/2^{n-1}} OA)).$$

By the choice of n, this difference should be very close to both x and 0, giving you a contradiction.)

Angle Measure Theorem 164 ([2] Axiom 7). *The degree measure function $\mathcal{D}$ has the following properties.*

(i) *The degree measure of a right angle is* 90.
(ii) *Angles are congruent if and only if they have the same degree measure.*
(iii) *If B is in the interior of $\angle AOC$, then*

$$\mathcal{D}(\angle AOC) = \mathcal{D}(\angle AOB) + \mathcal{D}(\angle BOC).$$

5.5 Consequences of Angle Measure

We end this chapter with some consequences of the Angle Measure Theorem 164.

Corollary 165. *If $\overrightarrow{OB}$ bisects $\angle AOC$, then*

$$\mathcal{D}(\angle AOB) = \mathcal{D}(\angle BOC) = \tfrac{1}{2}\mathcal{D}(\angle AOC).$$

Theorem 166 ([2] T82). *If $\triangle ABC$ is equilateral, then $\mathcal{D}(\angle A) = \mathcal{D}(\angle B) = \mathcal{D}(\angle C)$.*

Theorem 167 ([2] T86). *The sum of the degree measures of two supplementary angles is* 180.

Angle Sum Theorem 168 ([2] T87). *The sum of the degree measures of the three angles of a triangle is* 180.

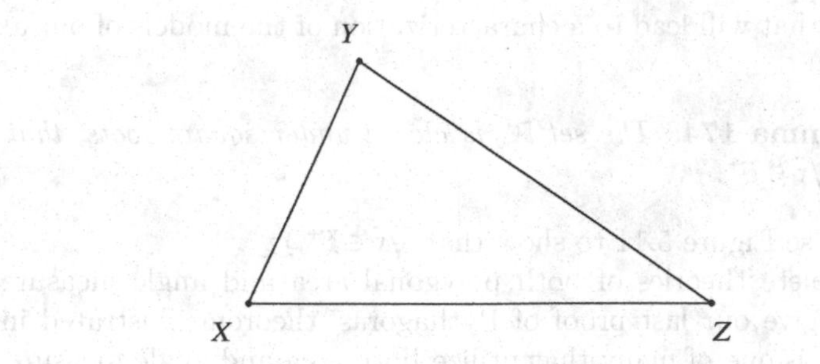

Fig. 5.10 $\mathcal{D}(\angle X) + \mathcal{D}(\angle Y) + \mathcal{D}(\angle Z) = 180.$

Corollary 169 ([2] C88). *Every angle of every equilateral triangle has degree measure* 60.

Corollary 170 ([2] C89). *The sum of the degree measures of the two non-right angles of a right triangle is* 90.

Corollary 171 ([2] C125). *Every pair of isosceles right triangles are similar and every pair of 30–60–90 triangles are similar.*

Strong Exterior Angle Theorem 172 ([2] T90). *The degree measure of an exterior angle of a triangle is equal to the sum of the degree measures of the two opposite interior angles.*

Theorem 173 ([2] T99). *If AB is a diameter of circle* **C** *with center O and C is a third point on* **C**, *then $\angle ACB$ is a right angle.*

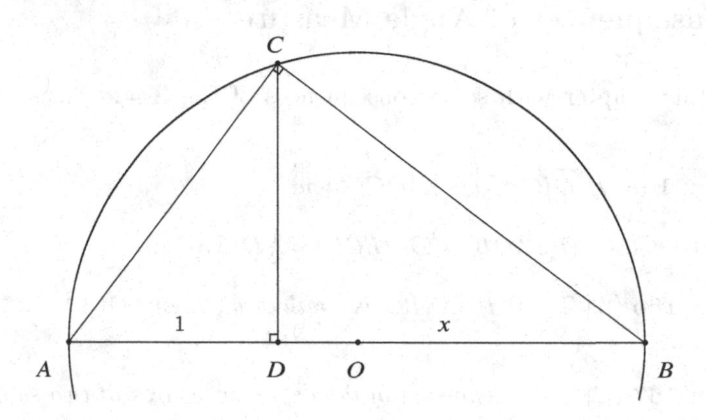

Fig. 5.11 $\sqrt{x} \in \mathbb{F}^+$.

As a final application of these ideas, you can prove a third and final closure lemma for $\mathbb{F}^+$ that will lead to a characterization of the models of our axioms in Chapter 8.

Third $\mathbb{F}^+$ Lemma 174. *The set $\mathbb{F}^+$ is closed under square roots, that is, if $x \in \mathbb{F}^+$, then $\sqrt{x} \in \mathbb{F}^+$.*

(Let $x \in \mathbb{F}^+$. Use Figure 5.11 to show that $\sqrt{x} \in \mathbb{F}^+$.)

With complete theories of both polygonal area and angle measure now available, we give our last proof of Pythagoras' theorem illustrated in Figure 5.12. This is one of many that utilize both area and angle measure.

5th Pythagorean Theorem 175 ([2] T92). *If a right triangle has legs of lengths a and b and hypotenuse of length c, then $c^2 = a^2 + b^2$.*

(Be sure to prove that the inner quadrilateral is indeed a square!)

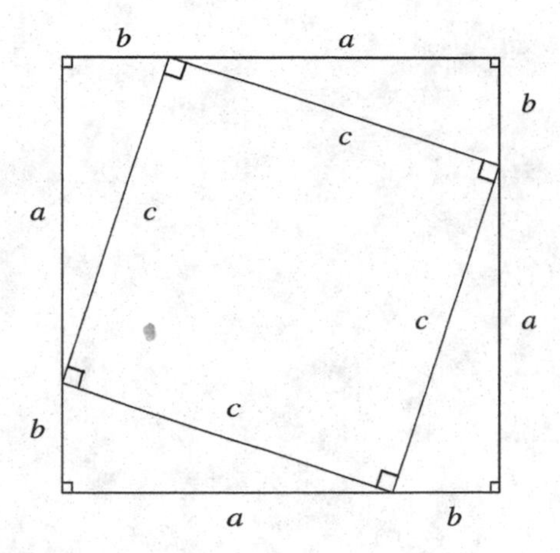

Fig. 5.12 Pythagorean Theorem 175. [2]

Chapter 6
Trigonometry

We say that an angle is **acute** if it is less than a right angle. This chapter is about that part of trigonometry that intersects high school geometry: the trigonometric functions of acute angles. We will start with the point at which these functions of acute angles arise from the study of similarity, and then establish some properties of them that will be needed in Chapters 7 and 8.

The degree measure of an acute angle is a real number in the open interval (0,90). We will denote by $\mathbb{D}$ the subset of (0,90) consisting of all numbers that are the degree measure of some acute angle. Our use of the set $\mathbb{D}$ contrasts with Birkhoff's Protractor Postulate, which says that *every* number in (0,90) is the degree measure of some acute angle. In Chapter 8 we will see that there are many models of our axioms in which $\mathbb{D}$ is only countably infinite, falling far short of being all of $(0, 90)$.

Lemma 176. *The following numbers are guaranteed to be in* $\mathbb{D}$.

(i) *If* $0 < i, n \in \mathbb{N}$ *and* $i < 2^{n-1}$, *then* $90i/2^{n-1} \in \mathbb{D}$.
(ii) *If* $x \in \mathbb{D}$, *then* $x/2 \in \mathbb{D}$.
(iii) *The numbers* 30 *and* 60 *are in* $\mathbb{D}$, *but not included in* (i).
(iv) *If* $x \in \mathbb{D}$ *and* $x < 45$, *then* $2x \in \mathbb{D}$.

If $x \in \mathbb{D}$, we label an angle in a figure with the symbol "$x°$" to indicate that it has degree measure x. Given a particular number $x \in \mathbb{D}$, Figure 6.1 illustrates several right triangles having an angle of measure x.

By Corollary 90 (A) we know that these triangles are all similar. By Theorem 88 we know that their corresponding pairs of sides are in the same ratio:

$$\frac{o}{a} = \frac{o'}{a'} = \frac{o''}{a''}, \quad \frac{o}{h} = \frac{o'}{h'} = \frac{o''}{h''}, \quad \frac{a}{h} = \frac{a'}{h'} = \frac{a''}{h''}.$$

This means that the values of these three ratios depend only on the measure x of the angle and not which triangle is used to calculate it. The same is true of the inverses of these ratios. Each of these ratios defines a real valued function on $\mathbb{D}$. For $x \in \mathbb{D}$ we define

D. M. Clark, S. Pathania, *A Full Axiomatic Development of High School Geometry*,
https://doi.org/10.1007/978-3-031-23525-2_6

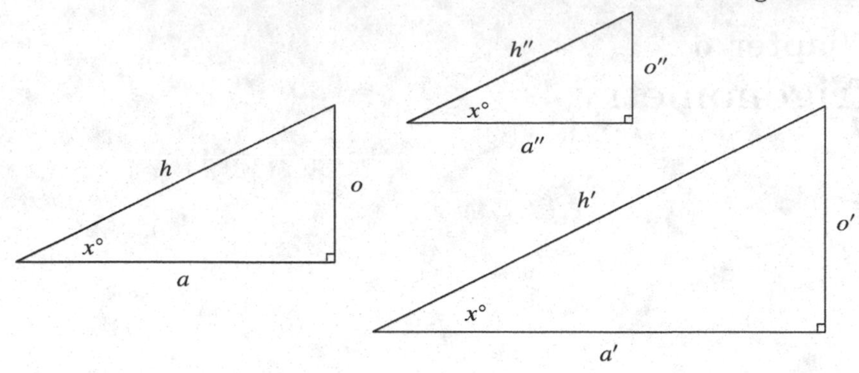

Fig. 6.1 Right triangles with an angle of measure x. [2]

$\tan(x) = \frac{o}{a}$, the **tangent** of x,
$\sin(x) = \frac{o}{h}$, the **sine** of x,
$\cos(x) = \frac{a}{h}$, the **cosine** of x,
$\cot(x) = \frac{a}{o}$, the **cotangent** of x,
$\csc(x) = \frac{h}{o}$, the **cosecant** of x,
$\sec(x) = \frac{h}{a}$, the **secant** of x.

Together these are called the **trigonometric functions**. By the 2nd $\mathbb{F}^+$ Lemma 70 each of these functions takes values in the set $\mathbb{F}^+$:

$$\tan, \sin, \cos, \cot, \csc, \sec : \mathbb{D} \to \mathbb{F}^+.$$

The first three are illustrated in Figure 6.2. If $\angle A$ is an angle, then "$\sin(\angle A)$", for example, means $\sin(\mathcal{D}(\angle A))$.

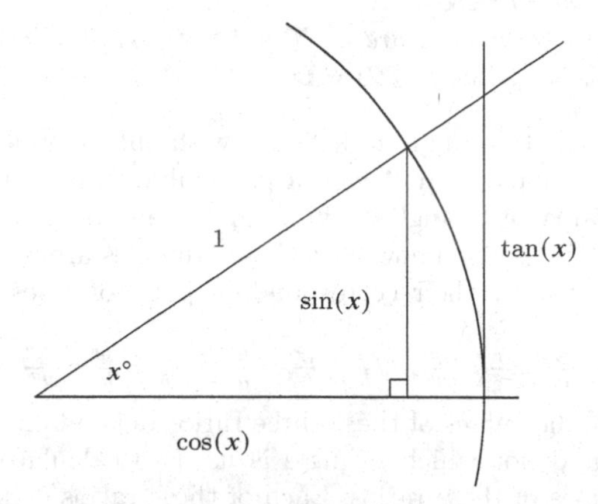

Fig. 6.2 Sine, cosine and tangent in a unit circle. [2]

At this point we would like to make some observations about the set $\mathbb{D}$ of acute angle measures. Figure 6.3 is an illustration of a triangle that may or may not exist in a particular plane. Specifically, such a triangle will exist if and only if a and b are in $\mathbb{F}^+$ and x is in $\mathbb{D}$. Thus such a triangle will be found in the plane for all choices of $a, b \in \mathbb{R}^+$ and all choices of $x \in (0, 90)$ if $\mathbb{F}^+ = \mathbb{R}^+$ and $\mathbb{D} = (0, 90)$. Recall that Birkhoff's Ruler Postulate (pages 9 and 100) implies that $\mathbb{F}^+ = \mathbb{R}^+$. Birkhoff used a similar assumption, his Protractor Postulate (page 101), which implies that $\mathbb{D} = (0, 90)$. Consequently, it is customary to add these two postulates in high school texts as further axioms so that students are guaranteed that $\mathbb{F}^+ = \mathbb{R}^+$ and $\mathbb{D} = (0, 90)$.

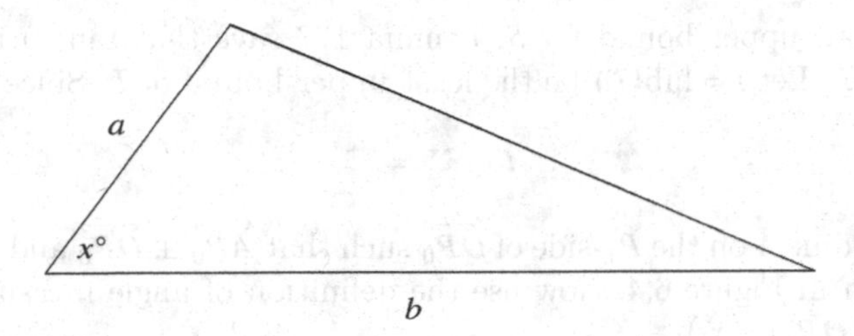

Fig. 6.3 A possible triangle?

We have not included either of these postulates for two reasons. One is that, as you are seeing, they are simply not needed to establish the standard theorems of Euclidean geometry. The other is that, in general, as we add more axioms to a mathematical theory we can prove more theorems at the cost of having fewer models to which they apply. In short, we have the choice of saying more about less or saying less about more.

In Chapter 8 you will see that this tradeoff is very stark in the present context. There we show that our axioms have a vast array of models, including uncountably many countable models. If we assume as additional axioms the Ruler Postulate and that our angle measure function $\mathcal{D}$ satisfies the Protractor Postulate, then we will be able to guarantee the existence of the triangle in Figure 6.3 for all $a, b \in \mathbb{R}^+$ and all $x \in (0, 90)$. But then Corollary 247 will tell us that the axioms will have only one model, the coordinate plane of the real numbers.

However, for the benefit of those who prefer to use Birkhoff's two postulates, we would like to point out something interesting. If you add the Ruler Postulate to our axioms, then you can prove that our angle measure function $\mathcal{D}$ satisfies the Protractor Postulate as a theorem and therefore not need to assume it as an axiom. To see how to do this, consider the setting in which $\mathbf{C}$ is a circle with center O, with radius 1 and with a set $\mathbf{X} \subseteq \mathbf{C}$ of binary points.

Lemma 177. *The tangent is an increasing function of angle measure, that is,* $\angle AOP_0 < \angle BOP_0$ *implies* $\tan(\angle AOP_0) < \tan(\angle BOP_0)$.

Theorem 178. *If every positive real number is the length of some segment, then every number between 0 and 90 is the degree measure of some acute angle. In symbols,*

$$\mathbb{F}^+ = \mathbb{R}^+ \textit{ implies } \mathbb{D} = (0, 90).$$

(Assume $\mathbb{F}^+ = \mathbb{R}^+$ and let $x \in (0, 90)$. You need to find an angle with degree measure x. Let

$$S := \left\{ \tfrac{90i}{2^{n-1}} < x \mid i, n \in \mathbb{Z}^+ \right\} \text{ and } T := \left\{ \tan\left(\tfrac{90i}{2^{n-1}}\right) \mid \tfrac{90i}{2^{n-1}} \in S \right\}.$$

Since x is an upper bound for S, Lemma 177 says that $\tan(x)$ is an upper bound for T. Let $t = \text{lub}(T)$ be the least upper bound of T. Since

$$t \in \mathbb{R}^+ = \mathbb{F}^+,$$

there is a point A on the P_1-side of $\overleftrightarrow{OP_0}$ such that $\overleftrightarrow{AP_0} \perp \overleftrightarrow{OP_0}$ and $\mathcal{L}(AP_0) = t$ as is shown in Figure 6.4. Now use the definition of angle measure to prove that $\mathcal{D}(\angle AOP_0) = x$.)

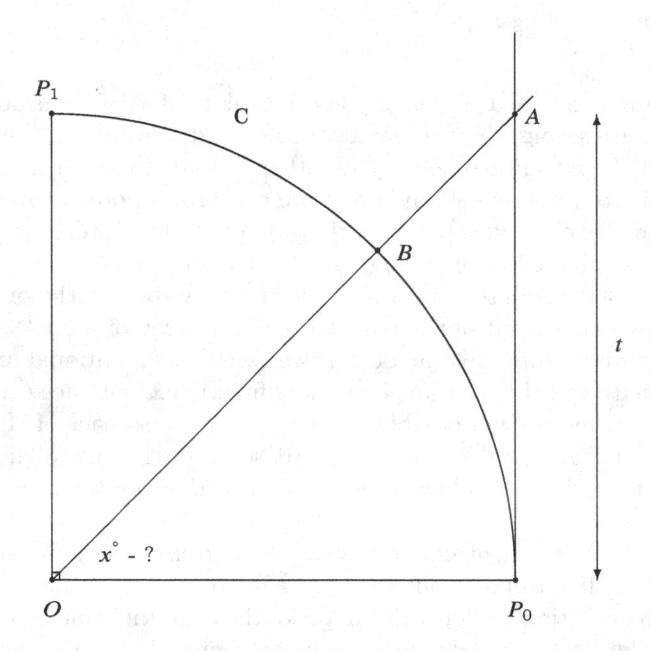

Fig. 6.4 Proof of Theorem 178.

In Chapter 2 we proved several theorems about right triangle congruence:

HL LL HA LA.

A right triangle is described by five measurements: two acute angles and three sides. Altogether these theorems say that if we make any two of these measurements that include at least one side, then there is only one possible outcome for each of the other three. This should raise an obvious question: Assuming there is a triangle with two of those measurements, how can we calculate the unique values of the other three?

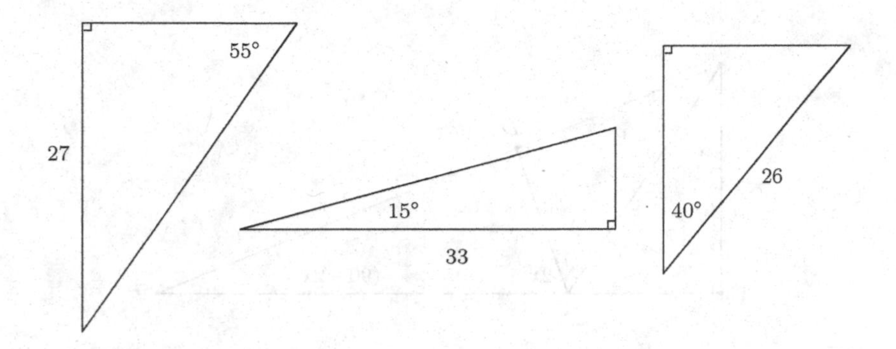

Fig. 6.5 Right triangles with three missing measurements. [2]

At this point high school students can answer this question by using trigonometric functions to find values of these missing measurements. For example, given a small table of trigonometric values or a calculator, they can approximate the values the three missing measurements in each of the triangles shown in Figure 6.5.

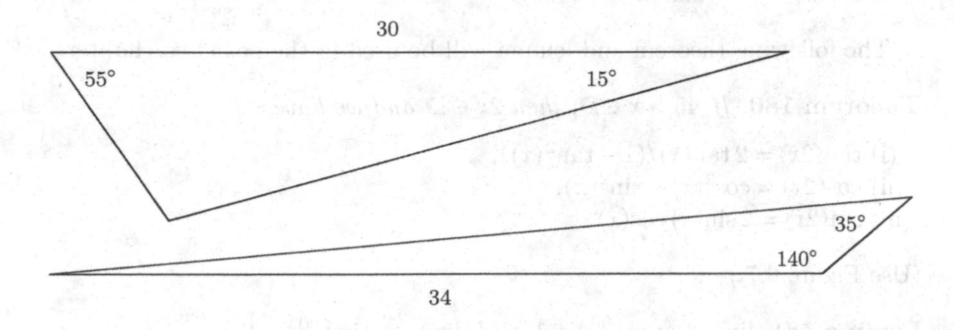

Fig. 6.6 Other triangles with three missing measurements. [2]

Doing indirect measurements in a triangle that is not a right triangle requires a little more ingenuity. Here we refer to our congruence axiom and theorems

$$\text{SSS} \quad \text{SAS} \quad \text{ASA} \quad \text{AAS}.$$

These are similar to the right triangle congruence theorems. Each of them tells us that if we make three particular direct measurements in a triangle, then there is only one possible value for each of the remaining three measurements. For example, ASA and AAS guarantee that the three missing measurements in each of the triangles in Figure 6.6 are uniquely determined. Using trigonometry high school students can calculate all of those values.

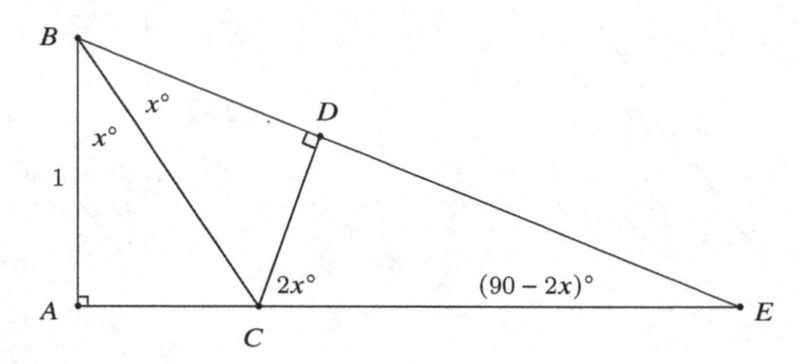

Fig. 6.7 $\tan(2x) = \frac{2\tan(x)}{1-\tan^2(x)}$

Lemma 179 ([2] T134). *For every* $x \in \mathbb{D}$ *we have*

(i) $\sin^2(x) + \cos^2(x) = 1$,
(ii) $\tan^2(x) + 1 = \sec^2(x)$,
(iii) $\sin(x) = \sqrt{1-\cos^2(x)}$,
(iv) $\cos(x) = \sqrt{1-\sin^2(x)}$.

The following theorem and lemma will be used in the next two chapters.

Theorem 180. *If* $45 > x \in \mathbb{D}$, *then* $2x \in \mathbb{D}$ *and we have*

(i) $\tan(2x) = 2\tan(x)/(1-\tan^2(x))$,
(ii) $\cos(2x) = \cos^2(x) - \sin^2(x)$,
(iii) $\sin(2x) = 2\sin(x)\cos(x)$.

(Use Figure 6.7.)

Lemma 181. $\lim_{n\to\infty} \cos\left(\frac{90}{2^n}\right) = 1$ *and* $\lim_{n\to\infty} \sin\left(\frac{90}{2^n}\right) = 0$.

(See Figure 6.8.)

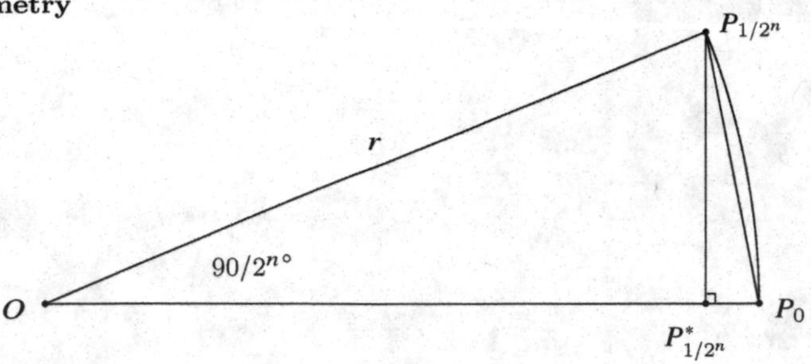

Fig. 6.8 Lemma 181.

Chapter 7
Circle Measure

In this chapter we will prove that there is a real number π such that, for every circle **C** with radius r, the circumference c and the area c are given by

$$c = 2\pi r \quad \text{and} \quad a = \pi r^2.$$

In the process we will show how to calculate the number π to as much accuracy as we like. To do this it will be necessary to define *circumference* in a way that is consistent with and in the spirit of the length measure function $\mathcal{L}$ and to define *area* in a way that is an extension of the area measure function $\mathcal{A}$. We will begin in Section 7.1 by extending area measure beyond polygonal regions to a broad range of other figures. By the **area** of circle **C** we will mean the area of the closed disc $\overline{\mathbf{C}}$ that is the union of **C** and its inside.

All of this will require finding convex polygons that are good approximations to the circle **C**. We say that a convex polygon is **inscribed** in **C** if its vertices are all on **C**. It **circumscribes C** if each of its sides is tangent to **C**. In Section 7.2 we will look for convex polygons that are inscribed in **C** and, in Section 7.3, convex polygons that circumscribe **C**. These approximations of **C** will be drawn directly from our work with angle measure in Chapter 5 and trigonometry in Chapter 6.

In the final Section 7.4 we will see if these approximations are good enough to apply the method of Section 7.1 for computing the area of **C** and to use a similar method to compute its circumference. The number π will appear along the way as a natural outcome of these investigations.

7.1 Jordan Measure of Area

There are many irregularly shaped figures that are not polygonal regions but can be closely approximated by polygonal regions, like the one illustrated in

D. M. Clark, S. Pathania, *A Full Axiomatic Development of High School Geometry*,
https://doi.org/10.1007/978-3-031-23525-2_7

Figure 7.1. In this section we will show how the domain of the area function $\mathcal{A}$ from Chapter 4 can be extended beyond polygonal regions to include these figures. If $\mathbf{Z}$ is a figure, let $\mathbb{L}$ be the set of all areas of polygonal regions contained in $\mathbf{Z}$ and let $\mathbb{U}$ be the set of all areas of polygonal regions that contain $\mathbf{Z}$. If there is exactly one positive real number x such that

$$a \leq x \leq b \text{ for all } a \in \mathbb{L} \text{ and } b \in \mathbb{U},$$

then we define $\mathcal{J}(\mathbf{Z})$, the **Jordan measure** [9] of $\mathbf{Z}$, to be x. Otherwise we say that $\mathbf{Z}$ is not Jordan measurable. Figure 7.1 illustrates an example of an irregularly shaped region $\mathbf{Z}$ and a polygonal region contained in it whose area $a \in \mathbb{L}$ is close to $\mathcal{J}(\mathbf{Z})$.

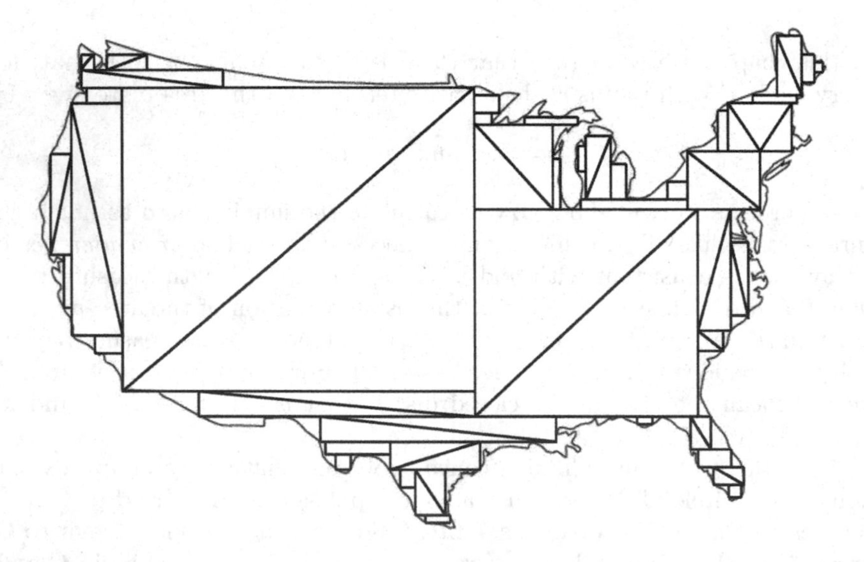

Fig. 7.1 Estimating an area with a polygonal region. [2]

At this point it is important to recognize that Jordan measure offers a conceptually satisfying notion of area. But it is not at all obvious that it agrees with the notion of area given in Chapter 4 when $\mathbf{Z}$ is a polygonal region. Also, the definition of Jordan measure gives us no practical way to calculate its value on a particular figure, as it asks us to examine the area of *every* polygonal region that is either contained in or contains that figure. Our goal in this section will be to prove two theorems. Theorem 200 will say that Jordan measure does indeed coincide with area measure when it is applied to polygonal regions. This fact will allow us to extend the notion of *area* by defining the area of an arbitrary figure to be its Jordan measure in case the figure is Jordan measurable. Theorem 201 will tell us how we can calculate

Jordan measure using only a carefully selected sample of these polygonal regions.

These two theorems will both be easy consequences of Theorem 199, whose proof will be the primary focus of our efforts in this section. As an immediate consequence of the definition of Jordan measure we find that, if **F** and **G** are Jordan measurable figures, then

$$\mathbf{F} \subseteq \mathbf{G} \text{ implies that } \mathcal{J}(\mathbf{F}) \leq \mathcal{J}(\mathbf{G}).$$

Theorem 199 will say that the area measure function $\mathcal{A}$ has the same property when **F** and **G** are restricted to polygonal regions:

$$\mathbf{F} \subseteq \mathbf{G} \text{ implies that } \mathcal{A}(\mathbf{F}) \leq \mathcal{A}(\mathbf{G}).$$

That assertion is the yet unproven part (iv) of the Area Measure Axiom of [2]. It may be helpful to skip to the end of this section first to see how the proofs of Theorems 200 and 201 both use nothing from this section other than Theorem 199.

As was the case with angle measure, a full treatment of area measure beyond polygonal area, and particularly to circles, is not included in the same three standard texts.

- Hartshorne's Chapter 5 in [7] gives a modern foundation for Euclid's work on areas. Variants of our Theorems 193 and 198 are given as exercises 22.1 and 22.2, but they are not used to extend area measure beyond polygonal regions.
- Millman and Parker [10] give a detailed account of polygonal area but make no mention of an extension to areas of other regions.
- Moise [11] presents Jordan measure to find the area of a circle by making direct use of Theorem 199 as we do. But he gives no reference or justification for this fact at all. (See [11], p336, Theorem 1 on p337 and Theorem 4 on p338.)

Estimating areas of irregular regions with polygonal regions is a practical application of geometry that is normally taught in high school and is included in [2]. (See, for example, Figure 7.17.) Theorem 199 provides the justification for this technique. Its proof will require that we step beyond classical geometry and draw on some techniques from topology. A few simple definitions, extending those from Section 4.1, will give us all the tools that we need to do this.

Let **F** be a figure and let P be a point. Then P is a **limit point** of **F** if every open disc containing P contains a point of **F** other than P. (Note that a limit point of **F** may or may not be in **F**.) With this terminology we see that the *boundary* $\mathbf{F}^b$ of **F** is just the set of points of **F** that are limit points of its complement and the *interior* $\mathbf{F}^i$ of **F** is the set of points of **F** that are not limit points of its complement. The **closure** of **F** is the set $\mathbf{F}^c$ consisting of **F** together will all of its limit points. We say that **F** is **closed** if it contains all of its limit points, that is, $\mathbf{F}^c = \mathbf{F}$. It is easy to see that the closure of every figure $\mathbf{F}$ is itself closed, that $\mathbf{F}^{cc} = \mathbf{F}^c$.

Adding to these standard operations on figures, we define the image of $\mathbf{F}$ under the composition $* = ic$ as the closure $\mathbf{F}^* := \mathbf{F}^{ic}$ of the interior of $\mathbf{F}$. Figure 7.2 shows how an application of $*$ can be thought of informally as shaving off whiskers and then regenerating bald spots. Note that

$$\text{if } \mathbf{F} \text{ is closed, then } \mathbf{F}^* \subseteq \mathbf{F}$$

since $\mathbf{F}^* = \mathbf{F}^{ic} \subseteq \mathbf{F}^c = \mathbf{F}$. We call $*$ the **grooming operation** and say that $\mathbf{F}$ is **well groomed** if grooming $\mathbf{F}$ with $*$ leaves it unchanged, that is, if $\mathbf{F}^* = \mathbf{F}$.

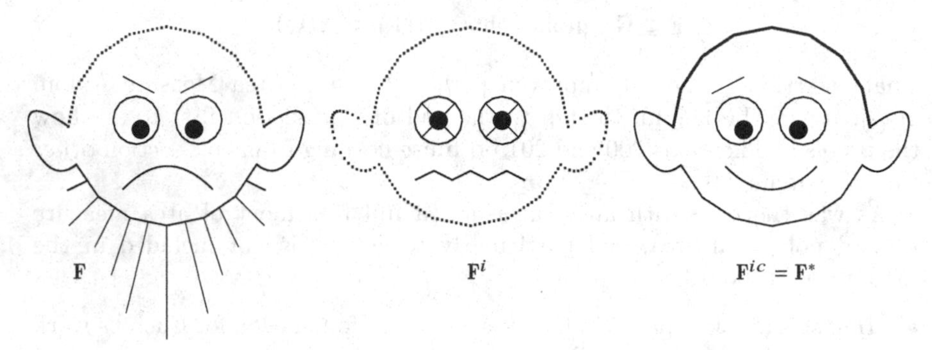

Fig. 7.2 The grooming operation $*$. (Here each figure includes all enclosed points.)

A slight reformulation of the definitions of these four operations on figures gives them useful uniform descriptions. (See Figure 7.3.)

Lemma 182. *Let P be a point and let $\mathbf{F}$ be a figure.*

(i) $P \in \mathbf{F}^b$ *if and only if* $P \in \mathbf{F}$ *and every open disc containing* P *intersects the complement of* $\mathbf{F}$.
(ii) $P \in \mathbf{F}^i$ *if and only if some open disc containing* P *is a subset of* $\mathbf{F}$.
(iii) $P \in \mathbf{F}^c$ *if and only if every open disc containing* P *intersects* $\mathbf{F}$.
(iv) $P \in \mathbf{F}^*$ *if and only if every open disc containing* P *intersects* $\mathbf{F}^i$.

The following lemma is an extension of Lemma 95 that will be often used in the topological arguments in this section.

Lemma 183. *If point P is in open disc $\mathbf{D}_1$ and in open disc $\mathbf{D}_2$, then P is also in an open disc $\mathbf{D}_3$ that is contained in $\mathbf{D}_1 \cap \mathbf{D}_2$ and has P as its center.*

Lemma 184. *A groomed figure $\mathbf{F}^*$ is well groomed, that is, $\mathbf{F}^{**} = \mathbf{F}^*$.*

Lemma 185. *Every open disc containing a point of a triangle $\mathbf{T}$ intersects the interior of $\overline{\mathbf{T}}$.*

Lemma 186. *Every triangular region is closed and well groomed.*

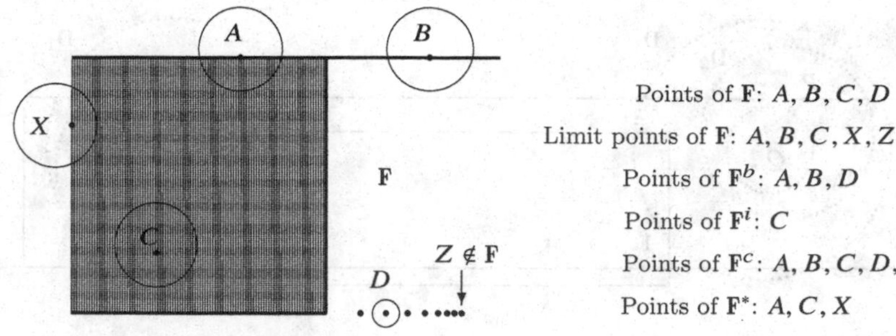

Fig. 7.3 Attributes of points.

In what follows we will use the grooming operation to create polygonal regions from figures that are not polygonal regions but are close to being polygonal regions. This process is illustrated in Figure 7.4.

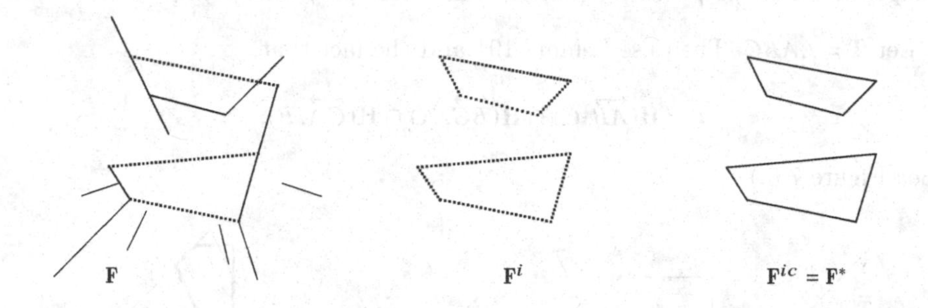

Fig. 7.4 Grooming a messy figure to make it into a polygonal region. (Again each figure includes all enclosed points.)

Lemma 187. *If at least one of the two closed figures* $\mathbf{F}$ *and* $\mathbf{G}$ *is well groomed, then* $(\mathbf{F} \cup \mathbf{G})^* = \mathbf{F}^* \cup \mathbf{G}^*$.

Corollary 188. *The union of a finite set of well groomed figures is well groomed.*

Corollary 189. *Every polygonal region is closed and well groomed.*

Lemma 190. *If* $\mathbf{F}$ *and* $\mathbf{G}$ *are well groomed, then* $(\mathbf{G} \backslash \mathbf{F})^* = (\mathbf{G} \backslash \mathbf{F})^c$.

(Assume $P \in (\mathbf{G} \backslash \mathbf{F})^c$. Consider two cases, $P \notin \mathbf{F}$ (Figure 7.5, left) and $P \in \mathbf{F}$ (Figure 7.5, right).)

Lemma 191. *The intersection of a half plane* $\mathbf{H}(\ell, A)$ *and a convex polygonal region* $\overline{\mathbf{X}}$ *is either empty, a single point, a segment or a convex polygonal region.*

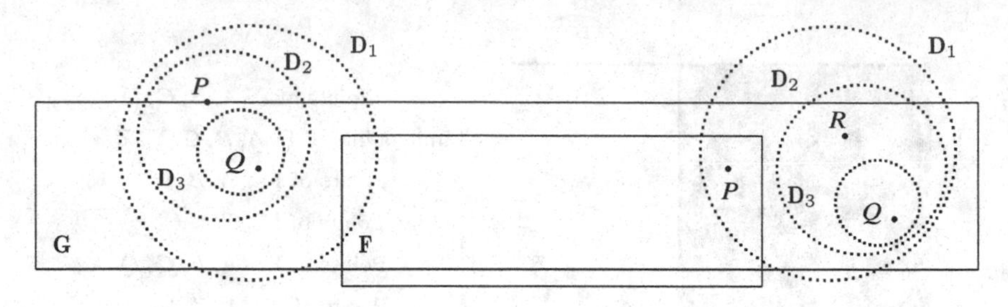

Fig. 7.5 $(\mathbf{G}\backslash\mathbf{F})^* \supseteq (\mathbf{G}\backslash\mathbf{F})^c$

(Assume the intersection contains two points and let Q be a point between them. Use Lemmas 94, 97 and 108.)

Lemma 192. *If* $\overline{\mathbf{S}}$ *and* $\overline{\mathbf{T}}$ *are triangular regions, then* $\overline{\mathbf{S}} \cap \overline{\mathbf{T}}$ *is either the empty set, a single point, a segment or a convex polygonal region.*

(Let $\mathbf{T} = \triangle ABC$. Then use Lemma 191 and the fact that

$$\overline{\mathbf{T}} = \mathbf{H}(\overleftrightarrow{AB}, C) \cap \mathbf{H}(\overleftrightarrow{BC}, A) \cap \mathbf{H}(\overleftrightarrow{CA}, B).$$

See Figure 7.6.)

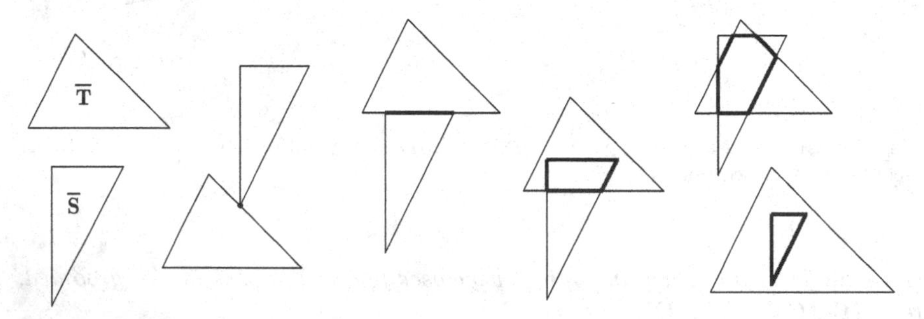

Fig. 7.6 Six samples of intersections of triangular regions, each outlined in boldface.

Based on Lemma 192 we might be tempted to conjecture that the intersection of two polygonal regions whose interiors intersect is again a polygonal region. However, the right column in Figure 7.9 shows that this is not true, but suggests that grooming the intersection might make it into a polygonal region.

Theorem 193. *If* **F** *and* **G** *are polygonal regions whose interiors intersect, then* $(\mathbf{F} \cap \mathbf{G})^*$ *is a polygonal region.*

(Let $\mathcal{U}$ and $\mathcal{V}$ be finite collections of triangles such that

$$\mathbf{F} = \bigcup\{\overline{\mathbf{S}} \mid \mathbf{S} \in \mathcal{U}\} \quad \text{and} \quad \mathbf{G} = \bigcup\{\overline{\mathbf{T}} \mid \mathbf{T} \in \mathcal{V}\}$$

are unions of triangular regions with disjoint interiors. Then we have

$$\begin{aligned}(\mathbf{F} \cap \mathbf{G})^* &= \left(\bigcup\{\overline{\mathbf{S}} \mid \mathbf{S} \in \mathcal{U}\} \cap \bigcup\{\overline{\mathbf{T}} \mid \mathbf{T} \in \mathcal{V}\}\right)^* \\ &= \left(\bigcup\{\overline{\mathbf{S}} \cap \overline{\mathbf{T}} \mid \mathbf{S} \in \mathcal{U} \text{ and } \mathbf{T} \in \mathcal{V}\}\right)^*.\end{aligned}$$

To determine what this is, we apply Lemma 192 and partition the regions $\overline{\mathbf{S}} \cap \overline{\mathbf{T}}$ into two collections, $\mathcal{W}$ and $\mathcal{W}'$, with $\mathcal{W}$ being those that are convex polygonal regions and $\mathcal{W}'$ being those that are empty, a point or a segment. Then we have

$$(\mathbf{F} \cap \mathbf{G})^* = \left(\bigcup \mathcal{W} \cup \bigcup \mathcal{W}'\right)^*.$$

Now show that this is a polygonal region using Theorem 99, Lemma 187, Corollary 188, Corollary 189 and Lemma 192.)

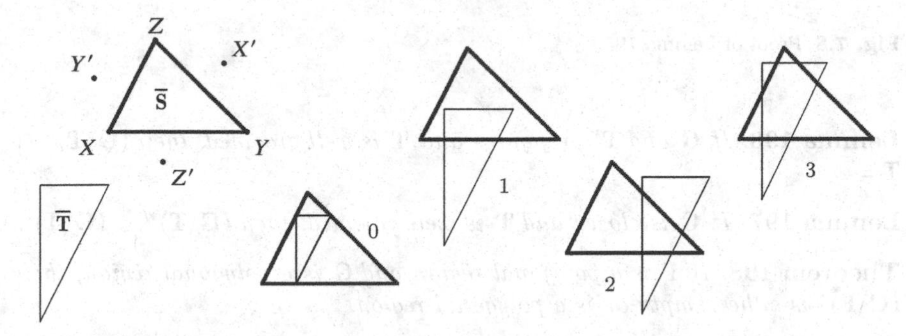

Fig. 7.7 Triangles **S** with 0, 1, 2 or 3 **T**-sides.

Theorem 194. *If* $\overline{\mathbf{S}}$ *and* $\overline{\mathbf{T}}$ *are triangular regions, then* $(\overline{\mathbf{T}} \backslash \overline{\mathbf{S}})^*$ *is either empty or is a polygonal region.*

(Let $\mathbf{S} = \triangle XYZ$. Choose points $X' \notin \mathbf{H}(\overleftrightarrow{YZ}, X)$, $Y' \notin \mathbf{H}(\overleftrightarrow{ZX}, Y)$ and $Z' \notin \mathbf{H}(\overleftrightarrow{XY}, Z)$ as in Figure 7.7. Let U, V and W be the three vertices of **S** in some order. Then UV is a **T**-*side* of **S** if **T** contains a point of $\mathbf{H}(\overleftrightarrow{UV}, W')$, that is, **T** protrudes beyond side UV of **S**.

Prove the theorem in four cases depending on the number of **T**-sides that **S** has. In each case label the vertices of **S** with "X", "Y" and "Z" in such a way that its **T**-sides precede its non-**T**-sides in the sequence XY, YZ, ZX. Use Lemmas 108, 191 and 192 to express $(\overline{\mathbf{T}} \backslash \overline{\mathbf{S}})^*$ as the union of polygonal regions with disjoint interiors. Then use Theorem 120(iii).)

Lemma 195. *If* $\overline{\mathbf{S}}$ *is a triangular region and* $\mathbf{G}$ *is a polygonal region, then* $(\mathbf{G} \backslash \overline{\mathbf{S}})^*$ *is either empty or is a polygonal region.*

(Assume $(\mathbf{G}\backslash\overline{\mathbf{S}})^* \neq \varnothing$ and consider a triangulation of $\mathbf{G}$. Let $\mathcal{U}$ be collection of those triangular regions of $\mathbf{G}$ not contained in $\overline{\mathbf{S}}$. Use previous results to verify these equalities:

$$(\mathbf{G}\backslash\overline{\mathbf{S}})^* = (\mathbf{G}\backslash\overline{\mathbf{S}})^c = \Big(\bigcup_{\overline{\mathbf{T}}\in\mathcal{U}} \overline{\mathbf{T}}\backslash\overline{\mathbf{S}}\Big)^c = \bigcup_{\overline{\mathbf{T}}\in\mathcal{U}} (\overline{\mathbf{T}}\backslash\overline{\mathbf{S}})^c = \bigcup_{\overline{\mathbf{T}}\in\mathcal{U}} (\overline{\mathbf{T}}\backslash\overline{\mathbf{S}})^*.)$$

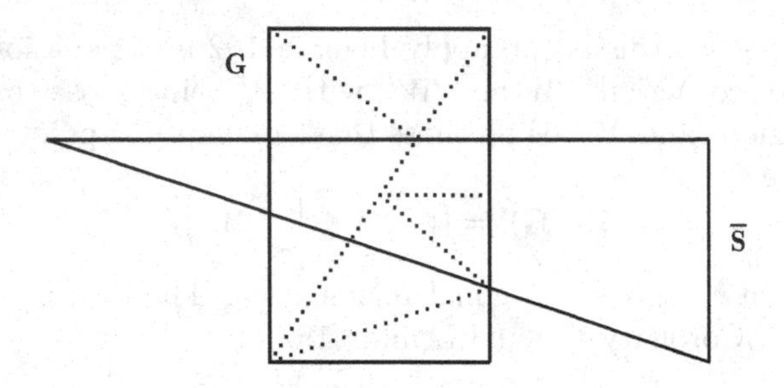

Fig. 7.8 Proof of Lemma 195.

Lemma 196. *If* $\mathbf{G}$ *and* $\mathbf{T}$ *are figures and* $\mathbf{T}$ *is well groomed, then* $(\mathbf{G}\backslash\mathbf{T})^{*i} \cap \mathbf{T} = \varnothing$.

Lemma 197. *If* $\mathbf{G}$ *is closed and* $\mathbf{T}$ *is well groomed, then* $(\mathbf{G}\backslash\mathbf{T})^{*i} \subseteq (\mathbf{G}\backslash\mathbf{T})^i$.

Theorem 198. *If* $\mathbf{F}$ *is a polygonal region and* $\mathbf{G}$ *is a polygonal region, then* $(\mathbf{G}\backslash\mathbf{F})^*$ *is either empty or is a polygonal region.*

(Use induction on the number n of triangular regions that make up $\mathbf{F}$, starting with Lemma 195. Assume it is true for n and let $\mathcal{U}$ be a collection of $n+1$ triangular regions with disjoint interiors that have $\mathbf{F}$ as their union. Pick one triangular region $\overline{\mathbf{T}} \in \mathcal{U}$ and let $\mathbf{H}$ be the union of the n remaining regions in $\mathcal{U}$. Then $\mathbf{F}$ and $\mathbf{H}$ have disjoint interiors, $\mathbf{F} = \overline{\mathbf{T}} \cup \mathbf{H}$ and $(\mathbf{G}\backslash\overline{\mathbf{T}})^*$ and $(\mathbf{G}\backslash\mathbf{H})^*$ are polygonal regions by induction. Now assume $(\mathbf{G}\backslash\mathbf{F})^* \neq \varnothing$. Then

$$\begin{aligned}\varnothing \neq (\mathbf{G}\backslash\mathbf{F})^i &\subseteq (\mathbf{G}\backslash\mathbf{H} \cap \mathbf{G}\backslash\overline{\mathbf{T}})^i \subseteq (\mathbf{G}\backslash\mathbf{H})^i \cap (\mathbf{G}\backslash\overline{\mathbf{T}})^i \\ &\subseteq (\mathbf{G}\backslash\mathbf{H})^{ici} \cap (\mathbf{G}\backslash\overline{\mathbf{T}})^{ici} = (\mathbf{G}\backslash\mathbf{H})^{*i} \cap (\mathbf{G}\backslash\overline{\mathbf{T}})^{*i}.\end{aligned}$$

Since $(\mathbf{G}\backslash\mathbf{H})^*$ and $(\mathbf{G}\backslash\overline{\mathbf{T}})^*$ are polygonal regions, $((\mathbf{G}\backslash\overline{\mathbf{T}})^* \cap (\mathbf{G}\backslash\mathbf{H})^*)^*$ is a polygonal region by Theorem 193. Complete the proof by showing that $(\mathbf{G}\backslash\mathbf{F})^* = ((\mathbf{G}\backslash\overline{\mathbf{T}})^* \cap (\mathbf{G}\backslash\mathbf{H})^*)^*$ as is shown in Figure 7.9.)

Using Theorem 198 and the Area Measure Theorem 120(iii) you can at last prove part (iv) of the Area Measure Axiom of [2].

Theorem 199. *If* $\mathbf{F}$ *and* $\mathbf{G}$ *are polygonal regions and* $\mathbf{F} \subseteq \mathbf{G}$, *then* $\mathcal{A}(\mathbf{F}) \leq \mathcal{A}(\mathbf{G})$.

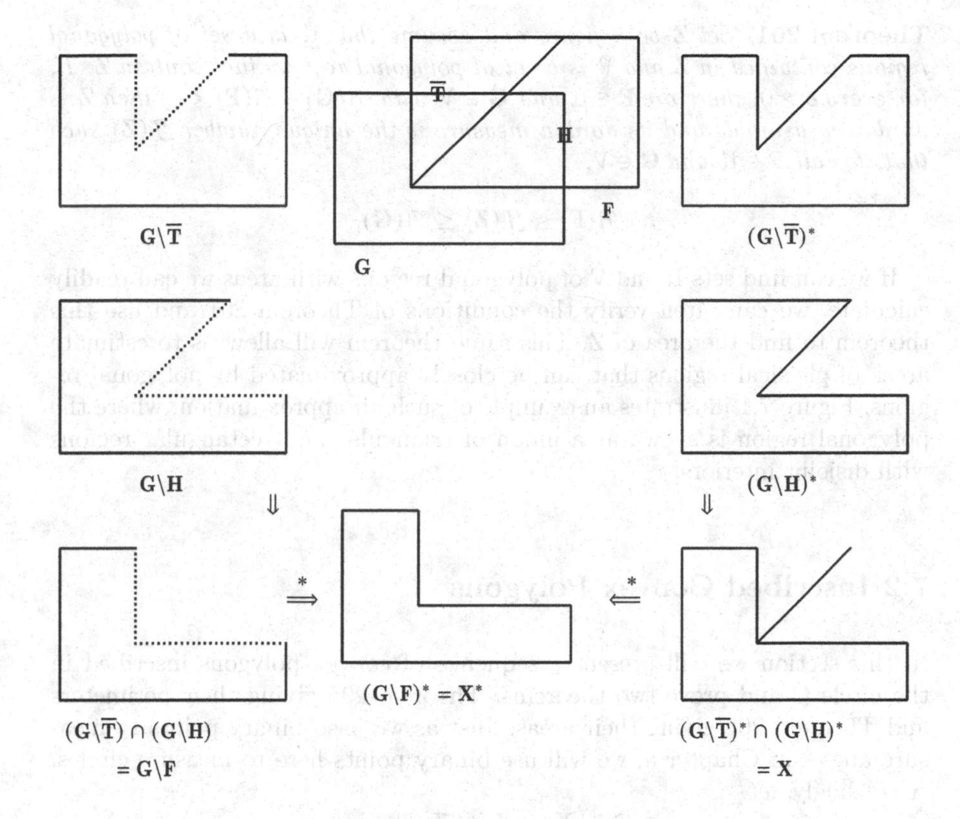

Fig. 7.9 Constructing $(\mathbf{G}\backslash\mathbf{F})^*$ and $((\mathbf{G}\backslash\overline{\mathbf{T}})^* \cap (\mathbf{G}\backslash\mathbf{H})^*)^*$.

As an immediate consequence of Theorem 199 we see that the Jordan measure function $\mathcal{J}$ is indeed an extension of the polygonal area measure function $\mathcal{A}$.

Theorem 200. *Every polygonal region* $\mathbf{Z}$ *is Jordan measurable, and its Jordan measure is the same as its area:*

$$\mathcal{J}(\mathbf{Z}) = \mathcal{A}(\mathbf{Z}).$$

Using this theorem we can unambiguously define the **area** of a Jordan measurable region $\mathbf{Z}$ as its Jordan measure $\mathcal{J}(\mathbf{Z})$.

As a second application of Theorem 199 and the completeness property of the real numbers, we have the following test for Jordan measurability and method of finding Jordan measure. This test and method were first introduced by Archimedes of Syracuse (c.287 BCE - 212 BCE) to find calculate the area of a circle. We will use Theorem 201 to give a modern version of that derivation.

Theorem 201. *Let* $\mathbf{Z}$ *be a figure and assume that* $\mathcal{U}$ *is a set of polygonal regions contained in* $\mathbf{Z}$ *and* $\mathcal{V}$ *is a set of polygonal regions that contain* $\mathbf{Z}$*. If, for every* $\epsilon > 0$*, there are* $\mathbf{F} \in \mathcal{U}$ *and* $\mathbf{G} \in \mathcal{V}$ *with* $\mathcal{A}(\mathbf{G}) - \mathcal{A}(\mathbf{F}) < \epsilon$*, then* $\mathbf{Z}$ *is Jordan measurable and its Jordan measure is the unique number* $\mathcal{J}(\mathbf{Z})$ *such that, for all* $\mathbf{F} \in \mathcal{U}$ *and* $\mathbf{G} \in \mathcal{V}$*,*

$$\mathcal{A}(\mathbf{F}) \leq \mathcal{J}(\mathbf{Z}) \leq \mathcal{A}(\mathbf{G})$$

If we can find sets $\mathcal{U}$ and $\mathcal{V}$ of polygonal regions with areas we can readily calculate, we can often verify the conditions of Theorem 201 and use this theorem to find the area of $\mathbf{Z}$. This same theorem will allow us to estimate areas of physical regions that can be closely approximated by polygonal regions. Figure 7.1 illustrates an example of such an approximation, where the polygonal region is shown as a union of triangular and rectangular regions with disjoint interiors.

7.2 Inscribed Convex Polygons

In this section we will present a sequence of convex polygons inscribed in the circle $\mathbf{C}$ and prove two theorems: Theorem 205 giving their perimeters and Theorem 208 giving their areas. Just as we used binary points to measure angles in Chapter 5, we will use binary points here to measure circles. Accordingly, let

$$\mathbf{X} = \mathbf{X}_1 \cup \mathbf{X}_2 \cup \mathbf{X}_3 \cup \ldots$$

be an arbitrary but fixed set of binary points of $\mathbf{C}$ as described in Chapter 5. For each positive integer n we define

$$\mathrm{i}_n := \bigcup \{P_{i/2^{n-1}} P_{(i+1)/2^{n-1}} \mid 0 \leq i < 2^{n+1}\}.$$

We will show that $\mathbf{I}_n$ is a convex polygon. (Figure 7.10.) If $0 \leq i < 2^{n+1}$, we refer to the angle $\angle P_{(i+1)/2^{n-1}} O P_{i/2^{n-1}}$ as a **central angle** of $\mathbf{I}_n$. Recall from Lemma 153 that every point not on one of these central angles is in the interior of exactly one of them.

Lemma 202. *If* n *is a positive integer and* $0 \leq i < 2^{n+1}$*, then no point of* $\mathbf{X}_n$ *is in the interior of* $\angle P_{(i+1)/2^{n-1}} O P_{i/2^{n-1}}$*.*

Lemma 203. *If* n *is a positive integer, then* $\mathbf{I}_n$ *is a convex polygon inscribed in* $\mathbf{C}$*.*

Figure 7.10 illustrates the convex polygons $\mathbf{I}_1, \mathbf{I}_2, \mathbf{I}_3, \ldots$ inscribed in $\mathbf{C}$. This illustration suggests that these convex polygons might be good inner approximations to the circle $\mathbf{C}$. To use this idea we define the **perimeter** of

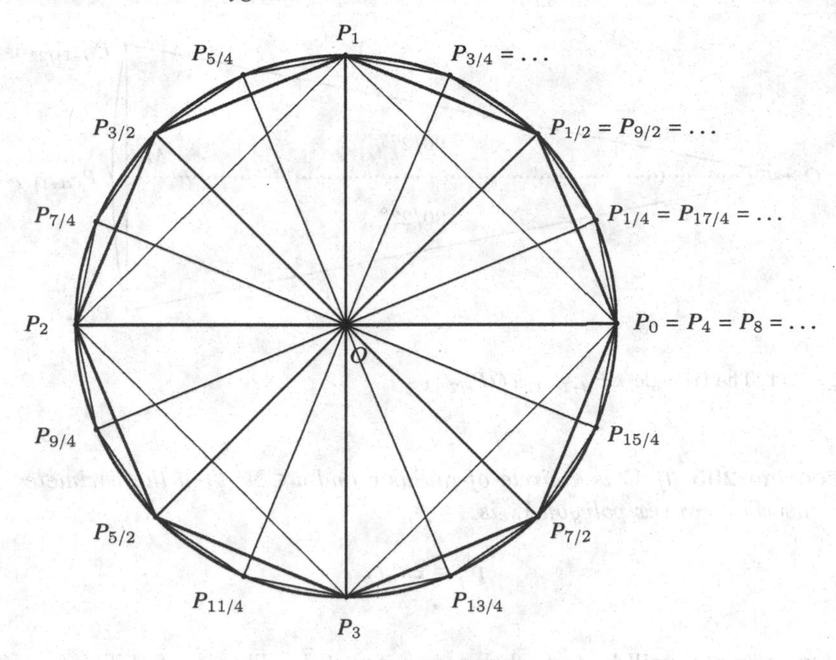

Fig. 7.10 Convex polygons $\bar{\mathbf{I}}_1, \bar{\mathbf{I}}_2, \bar{\mathbf{I}}_3$ inscribed in $\mathbf{C}$.

a convex polygon to be the sum of the lengths of its sides. Let $p_n^\flat$ denote the perimeter of the convex polygon $\mathbf{I}_n$:

$$p_n^\flat := \sum\{\mathcal{L}(P_{i/2^{n-1}}P_{(i+1)/2^{n-1}}) \mid 0 \le i < 2^{n+1}\}.$$

We can calculate the lengths of the sides of $\mathbf{I}_n$ by using the trigonometric functions from Chapter 6.

Lemma 204. *If circle* $\mathbf{C}$ *has radius* r *and* $n \in \mathbb{N}$, *then each side of the convex polygon* $\mathbf{I}_n$ *has length*

$$\mathcal{L}(P_{i/2^{n-1}}P_{(i+1)/2^{n-1}}) = 2r\sin\left(\tfrac{90}{2^n}\right).$$

Applying Lemma 204, we conclude that the perimeter of $\mathbf{I}_n$ is

$$p_n^\flat = 2^{n+1}\cdot 2r\sin\left(\tfrac{90}{2^n}\right) = 2^{n+2}r\sin\left(\tfrac{90}{2^n}\right)$$

for each positive integer n. It is helpful here to restate this conclusion by extracting the part of this quantity that is independent of the radius of the circle. For each positive integer n we define

$$\pi_n := 2^{n+1}\sin\left(\tfrac{90}{2^n}\right).$$

This gives us the following conclusion.

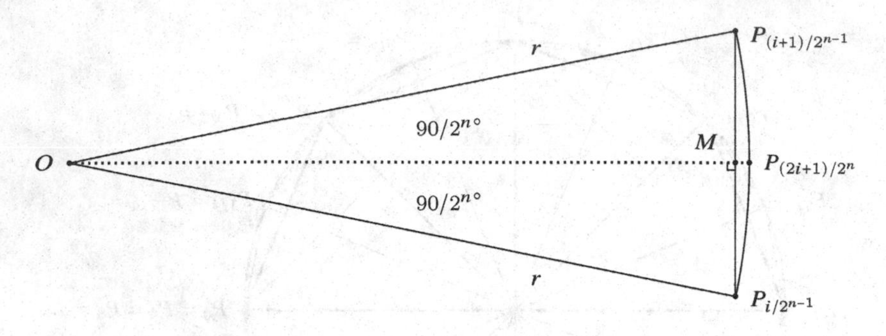

Fig. 7.11 The triangle $\Delta P_{(i+1)/2^{n-1}} O P_{i/2^{n-1}}$.

Theorem 205. *If* **C** *is a circle of radius* r *and* $n \in \mathbb{N}$, *then the perimeter of the inscribed convex polygon* $\mathbf{I}_n$ *is*

$$p_n^\flat = 2\pi_n r.$$

Our next step will be to find the area of each $\overline{\mathbf{I}}_n$. The proof of Theorem 99 shows that each $\overline{\mathbf{I}}_n$ is a convex polygonal region based on a standard triangulation for convex polygons. But our work here will be facilitated by using a different triangulation consisting of a set of congruent triangles. For each $n \in \mathbb{N}$ and $i < 2^{n+1}$ we will simplify notation with the abbreviation

$$\mathbf{T}_{i,n} := \Delta P_{(i+1)/2^{n-1}} O P_{i/2^{n-1}}.$$

Lemma 206. *If* $n \in \mathbb{N}$, *then the triangular regions* $\{\overline{\mathbf{T}}_{i,n} \mid i < 2^{n+1}\}$ *have disjoint interiors and*

$$\overline{\mathbf{I}}_n = \bigcup \{\overline{\mathbf{T}}_{i,n} \mid 0 \le i < 2^{n+1}\}.$$

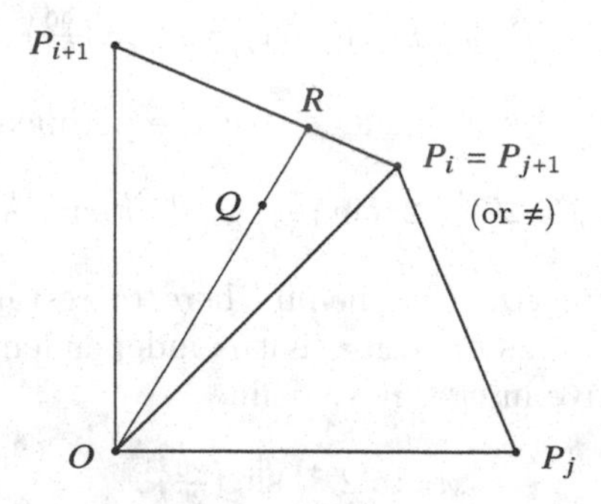

Fig. 7.12 Proof that $\overline{\mathbf{I}}_n \supseteq \bigcup \{\overline{\mathbf{T}}_{i,n} \mid 0 \le i < 2^{n+1}\}$.

Theorem 207. *For every positive integer n we have* $\overline{\mathbf{I}}_n \subseteq \overline{\mathbf{C}}$.

Theorem 208. *If* $\mathbf{C}$ *is a circle of radius r and n is a positive integer, then the area of the inscribed convex polygonal region* $\overline{\mathbf{I}}_n$ *is*

$$a_n^{\flat} = \pi_n r^2 \cos\left(\tfrac{90}{2^n}\right).$$

Together Theorems 207 and 208 tell us that, if $\overline{\mathbf{C}}$ is Jordan measurable, then its area is greater than $\pi_n r^2 \cos(90/2^n)$.

7.3 Circumscribed Convex Polygons

In this section we will present a sequence of convex polygons that circumscribe circle $\mathbf{C}$ with radius r and prove Theorem 210 giving their perimeters and areas. For each positive integer n, let $\mathbf{C}_n$ be the circle with center O and radius

$$r' := r \sec\left(\tfrac{90}{2^n}\right).$$

For larger values of n, the number $\sec(90/2^n)$ is only slightly bigger than one, and consequently $\mathbf{C}_n$ is a circle slightly larger than $\mathbf{C}$ with the same center O. For each $i < 2^{n+1}$, we take $Q_{i/2^{n-1}}$ to be the point of intersection of $\overrightarrow{OP_{i/2^{n-1}}}$ with $\mathbf{C}_n$. This gives us a set of binary points of $\mathbf{C}_n$ which, by Lemma 203, forms a convex polygon

$$\mathbf{O}_n := \bigcup\{Q_{i/2^{n-1}}Q_{(i+1)/2^{n-1}} \mid 0 \leq i < 2^{n+1}\}$$

that is inscribed in $\mathbf{C}_n$.

In contrast to the binary points $\{P_{i/2^{n-1}} \mid n \in \mathbb{N} \text{ and } i < 2^{n+1}\}$, all of which lie on the circle $\mathbf{C}$, we have here binary points $\{Q_{i/2^{n-1}} \mid n \in \mathbb{N} \text{ and } i < 2^{n+1}\}$ lying on a series of disjoint circles, namely,

$$\{Q_{i/2^{n-1}} \mid 0 \leq i < 2^{n+1}\} \subseteq \mathbf{C}_n.$$

(Figure 7.14.) When we write "$Q_{i/2^{n-1}}$", the denominator 2^{n-1} indicates that we mean the binary point on $\mathbf{C}_n$. Thus "$Q_{i/2^{n-1}}$" and "$Q_{2i/2^n}$" refer to different points because they lie on different circles.

For each $i < 2^{n+1}$, let $\mathbf{S}_{i,n} = \triangle Q_{i/2^{n-1}} O Q_{(i+1)/2^{n-1}}$. By Lemma 206 the polygonal region $\overline{\mathbf{O}}_n$ is the union of these triangular regions:

$$\overline{\mathbf{O}}_n = \bigcup\{\overline{\mathbf{S}}_{i,n} \mid 0 \leq i < 2^{n+1}\}.$$

Thus far we have not said anything about the relation between the convex polygon $\mathbf{O}_n$ and the original circle $\mathbf{C}$.

Theorem 209. *For every positive integer n we have* $\overline{\mathbf{C}} \subseteq \overline{\mathbf{O}}_n$.

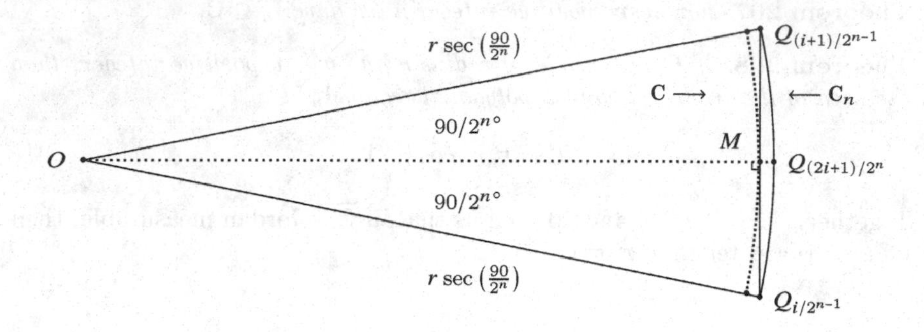

Fig. 7.13 $\mathbf{S}_{i,n} := \triangle Q_{(i+1)/2^{n-1}} O Q_{i/2^{n-1}}$ with $\mathbf{C}$ and $\mathbf{C}_n$. [2]

Theorem 210. *Let* $\mathbf{C}$ *be a circle of radius* r *and let* n *be a positive integer. Then* $\mathbf{O}_n$ *is a convex polygon circumscribing* $\mathbf{C}$ *with perimeter and area given by*

$$p_n^\sharp = 2\pi_n r \sec\left(\tfrac{90}{2^n}\right) \quad \textit{and} \quad a_n^\sharp = \pi_n r^2 \sec\left(\tfrac{90}{2^n}\right).$$

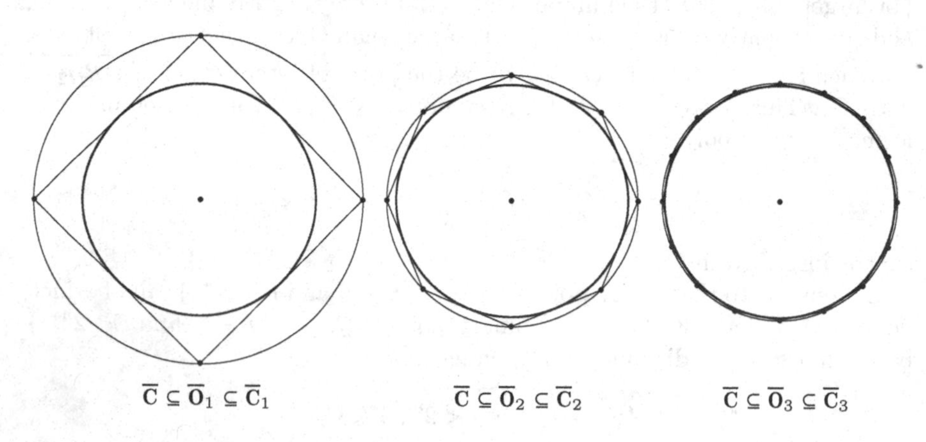

Fig. 7.14 First three upper approximations to circle $\mathbf{C}$.

Together Theorems 209 and 210 tell us that, if $\overline{\mathbf{C}}$ is Jordan measurable, then its area is less than $\pi_n r^2 \sec(90/2^n)$ for every positive integer n.

7.4 Area, Circumference and the Number π

We are now in a position to apply Theorem 201 to show that every circle is Jordan measurable, to calculate it area, and to define and calculate its circumference. These results will require that we first define the special number π and give a procedure to find its decimal digits.

To complete these tasks we will need to understand the asymptotic behavior of the sequences $[p_n^\flat]$, $[a_n^\flat]$, $[p_n^\sharp]$ and $[a_n^\sharp]$. Recall that

$$a_n^\flat := \mathcal{A}(\mathbf{I}_n) \text{ and } a_n^\sharp := \mathcal{A}(\mathbf{O}_n).$$

We have described these sequences in Theorems 205, 208 and 210 using the special sequence $[\pi_n]$, where

$$\pi_n := 2^{n+1} \sin\left(\tfrac{90}{2^n}\right).$$

Yet the asymptotic behavior of this sequence is not immediately apparent. The sequence $[2^{n+1}]$ rapidly approaches infinity while the sequence $[\sin(\frac{90}{2^n})]$ goes quickly to zero by Lemma 181. The answers to these questions start with an application of the Triangle Inequality.

Lemma 211. *The sequence $[p_n^\flat]$ is increasing.*

Theorem 212. *The sequence $[\pi_n]$ is increasing and bounded above by 4, and therefore converges to a number that we will denote by the Greek letter "π":*

$$\pi := \lim_{n\to\infty} \pi_n = \lim_{n\to\infty} 2^{n+1} \sin\left(\tfrac{90}{2^n}\right) \le 4.$$

In contrast to Lemma 211, Figure 7.14 suggests that the perimeters of the convex polygons $\mathbf{O}_n$ get progressively smaller. The proof of this fact is a direct application of the double angle formula of Theorem 180 for the tangent.

Lemma 213. *The sequence $[p_n^\sharp]$ is decreasing.*

Another sequence we need to inquire about is $[\pi_n \sec(90/2^n)]$. It is easy to see from Lemma 181 that this sequence must converge to π, but it is not easy to see at a glance whether it is either increasing or decreasing. The following lemma will have two important corollaries.

Lemma 214. *The sequence $[\pi_n \sec\left(\frac{90}{2^n}\right)]$ is decreasing.*

Corollary 215. *If n is a positive integer, then*

$$\pi_n < \pi < \pi_n \sec\left(\tfrac{90}{2^n}\right).$$

Notice that, while Theorem 212 gives a precise definition of the number π, it does not offer any constructive method to determine the decimal digits of π. It only tells us that π exceeds π_n for every $n \in \mathbb{N}$, and that $\pi \le 4$, but this information puts no constraint on the value of *any* decimal digit of π. Corollary 215 is significant because it gives us a direct way to calculate these digits. If the lower bound π_n and the upper bound $\pi_n \sec(90/2^n)$ round to the same first k digits, then those will be the first k digits of π. Calculating some initial windows like this shows that $[\pi_n]$ converges very rapidly. This process is illustrated in Figure 7.15 where we demonstrate that the correct value of π to eight decimal places is

$$\pi \approx 3.14159265$$

n	π_n	$< \pi <$	$\pi_n \sec(90/2^n)$
1	$2.82842712\cdots$	$< \pi <$	$4.00000000\ldots$
3	$3.12144515\cdots$	$< \pi <$	$3.18259787\ldots$
5	$3.14033115\cdots$	$< \pi <$	$3.14411838\ldots$
7	$3.14151380\cdots$	$< \pi <$	$3.14175036\ldots$
9	$3.14158772\cdots$	$< \pi <$	$3.14160251\ldots$
11	$3.14159234\cdots$	$< \pi <$	$3.14159326\ldots$
13	$3.14159263\cdots$	$< \pi <$	$3.14159269\ldots$
15	$3.14159265\cdots$	$< \pi <$	$3.141592653\ldots$

Fig. 7.15 Calculating the digits of π.

Our calculation of circle area is going to require knowing that the sequence $[a_n^\sharp]$ is decreasing. It is tempting to want to prove this using Theorem 199. But a close look at Figure 7.14 shows that $\overline{\mathbf{O}}_2 \not\subset \overline{\mathbf{O}}_1$ since every other vertex of $\overline{\mathbf{O}}_2$ lies outside of $\overline{\mathbf{O}}_1$. We can now prove this inequality without Theorem 199 by comparing our calculations of both values.

Corollary 216. *If n is a positive integer, then $a_{n+1}^\sharp < a_n^\sharp$.*

We now have all the pieces required to apply Theorem 201 to the circle $\mathbf{C}$. To do so we define

$$\mathcal{U} := \{\overline{\mathbf{I}}_n \mid 0 < n \in \mathbb{N}\} \text{ and } \mathcal{V} := \{\overline{\mathbf{O}}_n \mid 0 < n \in \mathbb{N}\}.$$

Corollary 217. *The sets $\mathcal{U}$ and $\mathcal{V}$ satisfy all of the conditions of Theorem 201.*

(i) *Each $\overline{\mathbf{I}}_n \in \mathcal{U}$ is a polygonal region contained in $\overline{\mathbf{C}}$.*
(ii) *Each $\overline{\mathbf{O}}_n \in \mathcal{V}$ is a polygonal region that contains $\overline{\mathbf{C}}$.*
(iii) *For each positive number ϵ, there is an $\overline{\mathbf{I}}_n \in \mathcal{U}$ and an $\overline{\mathbf{O}}_n \in \mathcal{V}$ such that $a_n^\sharp - a_n^\flat < \epsilon$.*

Applying Theorem 201 and Corollary 217 will give us two important consequences.

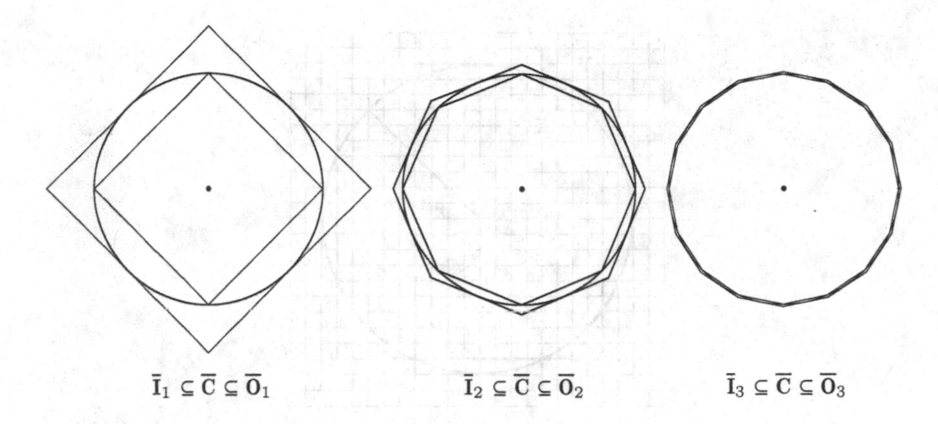

Fig. 7.16 $\bar{\mathbf{I}}_n \subseteq \overline{\mathbf{C}} \subseteq \overline{\mathbf{O}}_n$.

Theorem 218. *For every circle* $\mathbf{C}$, *the closed disc* $\overline{\mathbf{C}}$ *is Jordan measurable, and its Jordan measure is the unique number a such that*

$$a_n^\flat \leq a \leq a_n^\sharp$$

for every positive integer n.

The fact that $\overline{\mathbf{C}}$ is Jordan measurable alone provides a justification for estimating the area of a circle without having a formula for the area that is used in [2] and is taught in high schools. We place over the circle $\mathbf{C}$ of radius 10 a grid of one-by-one squares as shown in Figure 7.17. We then outline the polygonal region $\mathbf{X}$ consisting of all squares contained in $\overline{\mathbf{C}}$ and the polygonal region $\mathbf{Y}$ consisting of all squares that intersect $\overline{\mathbf{C}}$. The area of each of these regions is just the number of squares it contains. In the example shown in Figure 7.17, we conclude that

$$\mathcal{A}(\mathbf{X}) = 268 \leq a \leq 332 = \mathcal{A}(\mathbf{Y}).$$

Averaging these two bounds gives us an approximate estimate of the area as $a \approx 300$.

Ideally we would like to find the exact value of the area of $\mathbf{C}$. Theorem 201 tells us that, if we can find a particular number which is between $a_n^\flat$ and $a_n^\sharp$ for every $n \in \mathbb{N}$, then that number would have to be the exact area of $\mathbf{C}$.

Theorem 219. *The area of a circle* $\mathbf{C}$ *with radius r is*

$$a = \pi r^2.$$

This shows that the estimate $a \approx 300$ for the area of the circle of radius $r = 10$ in Figure 7.17 is not too far from the actual value $a = \pi 10^2 = 314.159\ldots$.

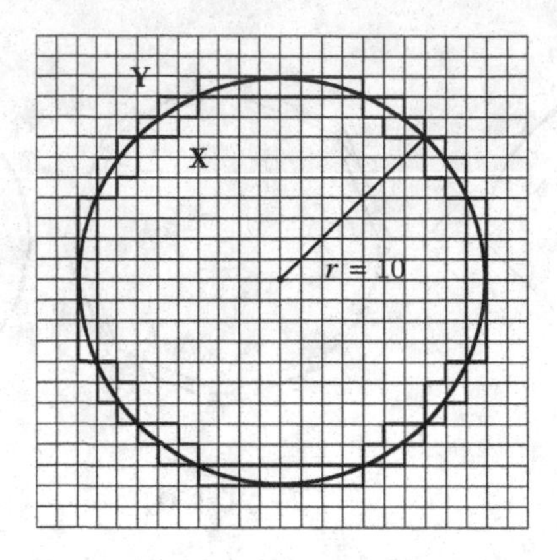

Fig. 7.17 Estimating the area of a circle. [2]

Figure 7.16 suggests that the perimeters $p_n^\flat$ and $p_n^\sharp$ rapidly approach each other as n grows large. This observation gives us a natural way to draw on the idea of Jordan measure in order to define the *circumference* of a circle. We will say that the circle $\mathbf{C}$ has **circumference** c if c is the unique number that is greater than the perimeter $p_n^\flat$ of $\mathbf{I}_n$ and less that the perimeter $p_n^\sharp$ of $\mathbf{O}_n$ for every positive integer n. The facts we now have show that every circle does have a circumference and say just what that circumference is.

Theorem 220. *The circumference of a circle* $\mathbf{C}$ *with radius* r *is*

$$c = 2\pi r.$$

Chapter 8
Consistency and Models

Chapter 7 concludes the investigation we laid out in the introduction to Chapter 1. We have stated ten axioms of geometry and proved a vast array of theorems from them that includes all of the standard facts of geometry that arise in the high school curriculum. This tells us that all of these theorems are true in every model $\mathcal{P} = \langle \mathbf{P}; d \rangle$ of our axioms. This sounds marvelous. However, before we celebrate, there is an important question we need to answer:

Is there really a plane $\mathcal{P}$ that is a model of our axioms?

This is by no means an idle question. The foundations of modern set theory and logic were first laid out in 1903 by Gottlob Frege [5], only to have this initial version undermined by Bertrand Russell whose famous paradox was obtained by proving a contradiction from Frege's axioms. When a contradiction is proven from a set of axioms, we say that the axioms are **inconsistent**. Since a contradiction cannot be true in any model, this means that there is no model of those axioms. As a result the theory built from those axioms is literally a theory about nothing. Before we conclude, we would like to find some reassurance that the axioms we have used here, and therefore also in [2], are indeed **consistent**, that is, no contradiction can be proven from them. The standard way to show that a set of axioms is consistent is to exhibit a model of those axioms.

If we can do this, we would ideally like to answer a second question:

Exactly which planes are models of our axioms?

This question is generally more difficult to answer. For example, the group axioms are consistent since every group is a model of them. But we are far from having a complete description of all groups.

In this chapter we will give a full answer to both of these questions for our axioms. Doing this requires an important shift in perspective. We will not be proving any more theorems *from* our axioms. Instead we will be proving theorems *about* our axioms. The statement that a set of axioms is or is not

D. M. Clark, S. Pathania, *A Full Axiomatic Development of High School Geometry*,
https://doi.org/10.1007/978-3-031-23525-2_8

consistent, for example, is a statement *about* those axioms. See Appendix B.10 for the connection of this work to [2].

8.1 Birkhoff's Geometry

To answer these questions we will draw on the text [10] *Geometry: A Metric Approach with Models* by Millman and Parker. This book offers a version of the metric axioms of Birkhoff [1] to provide a very clean axiomatic development of Euclidean and hyperbolic geometry that includes much of the what we have given here. We will present a brief outline of their development. All page references in this chapter refer to [10] and we will use some results cited from [10] as well.

Millman and Parker study what we will call an **MP-plane**, that is, a structure

$$\mathcal{G} = \langle \mathbf{P}; d, \mathcal{L}, m \rangle$$

where **P** is a set of objects called **points** that is endowed with three additional primitive constructs:

- $d : \mathbf{P} \times \mathbf{P} \to \mathbb{R}$, a **distance function** as we have used,
- $\mathcal{L}$, a non-empty collection of subsets of **P** called **lines**,
- m, a real valued function on subsets of **P** called **angles**.

Using just the primitive constructs of *point, distance* and *line*, the following constructs are defined in [10] for $\mathcal{G}$ exactly as we have defined them in this text:

between, segment, length, ray, circle,
radius, inside and outside of a circle.

The text [10] includes two very strong axioms of Birkhoff [1] that we have not used. The first says that every line can be coordinatized with the real numbers (pp 25-28).

Ruler Postulate. *For every line ℓ of $\mathcal{G}$ there is a bijection $f_\ell : \ell \to \mathbb{R}$ such that, for all $P, Q \in \ell$, we have $|f_\ell(P) - f_\ell(Q)| = d(P, Q)$.*

An MP-plane $\mathcal{G}$ is defined in [10] to be a *Euclidean geometry* if it satisfies a list of axioms that we now describe.

- An MP-plane is an *incidence geometry* if it satisfies the 2PA and the 3PA (p20).
- An incidence geometry is a *metric geometry* if it satisfies the Ruler Postulate (p25, p28).

If the Ruler Postulate holds in $\mathcal{G}$ and t is any positive real number, the surjectivity of f_ℓ tells us that there are points P and Q in ℓ such that $f_\ell(P) = t$

and $f_\ell(Q) = 0$. Thus $d(P,Q) = |t-0| = t$, and therefore the additive semigroup $\mathbb{F}^+$ associated with $\mathcal{G}$ consists of all positive real numbers, that is, $\mathbb{F}^+ = \mathbb{R}^+$.

In item #15, page 54, it is proven from these axioms that a subset ℓ of $\mathbf{P}$ is a **line** in $\mathcal{L}$ if and only if there are two points A and B with

$$\ell = \overleftarrow{AB} \cup \overrightarrow{AB}.$$

Consequently we can add more to our list of terms and phrases that have the same meaning in all models of the MP-axioms as they do in our geometry:

line, parallel lines, collinear, angle, triangle, sides of a line, interior of an angle, inside of a triangle, triangular region.

- A metric geometry is a *Pasch geometry* if it satisfies a condition that is equivalent to the PSA by Theorems 4.3.1 and 4.3.3 of [10].

The second axiom of [10] introduced by Birkhoff that we have not used concerns angle measure (p83).

Protractor Postulate. *If $\overrightarrow{BC}$ is a ray and t is a real number such that $0 < t < 180$, then on each side of $\overleftrightarrow{BC}$ there is a unique ray $\overrightarrow{BA}$ such that $m(\angle ABC) = t$.*

The function m on the angles of $\mathcal{G}$ is called a **degree measure function** if it satisfies the Protractor Postulate, the additivity condition (iii) of Theorem 164 and each angle has measure between 0 and 180.

- A *protractor geometry* is a Pasch geometry in which m is a degree measure function (p83, p84).

In a protractor geometry an angle in $\mathcal{G}$ is defined to be a **right angle** if it has measure 90. Two lines are **perpendicular** if they intersect at right angles. Angles are defined to be **congruent** if they have the same measure. (pp 96 -100.) We will need only one fact from [10] about angle measure.

Theorem 11.1.14. *Isometries preserve angle measure m.*

- A *neutral geometry* is a protractor geometry satisfying SAS (p118).
- A *Euclidean geometry* is a neutral geometry satisfying EUC (p183).

To summarize, a **Euclidean geometry** is an MP-plane in which d is a distance function, m is a degree measure function, our axioms 2PA, 3PA, PSA, SAS and EUC all hold and the Ruler Postulate holds. We will refer to this set of axioms for a Euclidean geometry as the **MP-axioms**.

Millman and Parker answer both of the questions we posed in the introduction of this chapter for the MP-axioms. To prove consistency, they take $\mathbb{R}$ to be the set of real numbers and form its coordinate plane $\mathcal{R}$ as follows. The points of $\mathcal{R}$ are ordered pairs (x, y) of numbers $x, y \in \mathbb{R}$ forming the set

$\mathbb{R}^2 := \mathbb{R}\times\mathbb{R}$. The **distance function** $d^{\mathbb{R}}$ is defined for $A = (x, y)$ and $B = (u, v)$ as

$$d^{\mathbb{R}}(A, B) := \sqrt{(x-u)^2 + (y-v)^2}.$$

The set $\mathcal{L}^{\mathbb{R}}$ consists of all **lines** $\overleftrightarrow{AB} := \overleftarrow{AB} \cup \overrightarrow{AB}$ where A and B are two points of $\mathbb{R}^2$. The **degree measure function** is defined using inner products, vector norms and the inverse cosine function as

$$m^{\mathbb{R}}(\angle BOA) := \cos^{-1}\left(\frac{\langle B-O, A-O\rangle}{||B-O||\cdot||A-O||}\right).$$

As they present the MP-axioms, they show progressively that each one holds in the MP-plane

$$\mathcal{R} = \langle \mathbb{R}^2; d^{\mathbb{R}}, \mathcal{L}^{\mathbb{R}}, m^{\mathbb{R}}\rangle.$$

This leads to an answer to our first question for the MP-axioms.

Theorem 221 (Problem 5, page 195 of [10]). *The MP-plane $\mathcal{R}$ is a model of the MP-axioms, which are therefore consistent.*

To outline the description in [10] of all models of the MP-axioms, we need an additional notion. We say that MP-planes

$$\mathcal{G} = \langle \mathbf{P}; d, \mathcal{L}, m\rangle \quad\text{and}\quad \mathcal{G}^* = \langle \mathbf{P}^*; d^*, \mathcal{L}^*, m^*\rangle$$

are **MP-isomorphic** if there is a bijection $\varphi : \mathbf{P} \to \mathbf{P}^*$, called an **MP-isomorphism**, that preserves the three MP-primitive constructs, that is,

- $d(A, B) = d^*(A^\varphi, B^\varphi)$ for all $A, B \in \mathbf{P}$,
- a set $\ell \subseteq \mathbf{P}$ is in $\mathcal{L}$ if and only if $\ell^\varphi \subseteq \mathbf{P}^*$ is in $\mathcal{L}^*$ and
- $m(\angle ABC) = m^*(\angle A^\varphi B^\varphi C^\varphi)$ for all angles $\angle ABC$ of $\mathcal{G}$.

We think of MP-planes that are MP-isomorphic as being essentially the same. In order to describe all models of the MP-axioms, we would like to identify a set $\mathcal{M}$ of models with the property that no two models in $\mathcal{M}$ are isomorphic to each other and every model is isomorphic to some model in $\mathcal{M}$.

We can identify MP-isomorphisms by looking at a simpler notion. An **isometry** from an MP-plane $\mathcal{G}$ to an MP-plane $\mathcal{G}^*$ is defined in [10], page 275, to be a function $\varphi : \mathbf{P} \to \mathbf{P}^*$ that preserves distance, that is,

$$d(A, B) = d^*(A^\varphi, B^\varphi)$$

for all $A, B \in \mathbf{P}$. It turns out that preservation of distance guarantees preservation of lines and angle measure as well.

Theorem 222 (Theorem 11.1.14 and Lemma 11.1.17 of [10]). *Every isometry from one model of the MP-axioms to another is an MP-isomorphism.*

Millman and Parker answer our second question for the MP-axioms by taking $\mathcal{M} = \{\mathcal{R}\}$, that is, by showing that $\mathcal{R}$ is essentially the only model of the MP-axioms. In general a set of axioms is said to be **categorical** if it has, up to isomorphism, only one model.

Lemma 223 (Theorem 11.1.20 of [10]). *Let* $\mathcal{G} = \langle \mathbf{P}; d, \mathcal{L}, m\rangle$ *be a model of the MP-axioms, let* O *be a point of* $\mathbf{P}$ *and let* ℓ_1 *and* ℓ_2 *be two lines which are perpendicular at* O*. Choose* $B \in \ell_1$ *and* $C \in \ell_2$ *so that* $B \neq O \neq C$*. Then there is an isometry* φ *from* $\mathcal{G}$ *to* $\mathcal{R}$ *such that* $O^\varphi = (0,0)$*, that* $B^\varphi = (x,0)$ *and* $C^\varphi = (0,y)$ *for some positive* $x, y \in \mathbb{R}$*.*

Theorem 224. *The MP-axioms are categorical with* $\mathcal{R}$ *as a model that is MP-isomorphic to every other model.*

8.2 Consistency

We will use the consistency result of Theorem 221 to establish the consistency of our axioms. Let

$$\mathcal{G} = \langle \mathbf{P}; d, \mathcal{L}, m\rangle$$

be any model of the MP-axioms. We will show that the plane

$$\mathcal{G}' := \langle \mathbf{P}; d\rangle$$

with the same set $\mathbf{P}$ of points and the same distance function d is a model of our axioms. Since $\mathcal{R}$ is a model of the MP-axioms, it will follow that our axioms are consistent with $\mathcal{R}'$ as a model.

Having the distance function d on the set $\mathbf{P}$ of points, the definitions of the defined constructs in Chapters 1 through 7 immediately give us a meaning of each of those defined constructs in $\mathcal{G}'$. In order to show that one of our axioms holds in $\mathcal{G}'$, we will first need to verify that the constructs mentioned in that axiom all have the same meanings in $\mathcal{G}'$ as they have in $\mathcal{G}$. It will then follow that the axiom has the same meaning in both. If we can then show that the axiom is provably true in $\mathcal{G}$ using the MP-axioms, it will follow that it also holds in $\mathcal{G}'$.

We begin doing this by observing that *between* has exactly the same definition in $\mathcal{G}'$ as in $\mathcal{G}$. It follows that all terms we found in the previous section to be defined using only *point*, *distance* and *between* in this text exactly as they are defined in [10] must also have the same meaning in $\mathcal{G}'$ as they have in $\mathcal{G}$. We list these terms together as

segment, ray, length, $\mathbb{F}^+$*, circle, radius, inside and outside of a circle, line, collinear, parallel lines, angle, triangle, sides of a line, interior of an angle, inside of a triangle, triangular region.*

This means, for example, that the same subsets of $\mathbf{P}$ are segments, rays or triangles in $\mathcal{G}$ and in $\mathcal{G}'$. A triangle has the same inside in $\mathcal{G}$ as in $\mathcal{G}'$. The same numbers are lengths of segments in $\mathcal{G}$ and in $\mathcal{G}'$. Once we establish that a term or phrase has the same meaning in $\mathcal{G}$ as it has in $\mathcal{G}'$, we will use it freely without needing to specify which plane we intend. If an axiom or theorem uses only terms that have the same meaning in both planes, then it is true in one plane if and only if it is true in the other. A number of our axioms use only *distance, between, length* and the above other constructs that have the same meaning in both planes. They give us a running start on what we need to establish.

Lemma 225. *The following axioms hold in* $\mathcal{G}'$*:* 2PA, 3PA, UNQ, INF, PSA, RCA, EUC.

It remains to show that our other three axioms, CCA, LMA and SSS, also hold in $\mathcal{G}'$. The CCA is not explicitly stated in [10] but can be proven from results in [10].

Lemma 226. *The* CCA *holds in* $\mathcal{G}'$.

(The facts established in our Section 5.1 are all proven from the MP-axioms in [10], as are both of the following.

- Problem 10, page 159. Between any point inside a circle and any point outside the circle there is a point on the circle.
- Problem 3, page 165. If two circles each contain a point inside the other, then they intersect in exactly two points.)

Because the final axioms LMA and SSS both reference isometries, we will need the following lemma.

Lemma 227. *A function* $\varphi : \mathbf{P} \to \mathbf{P}$ *is an isometry in* $\mathcal{G}'$ *if and only if it is an isometry in* $\mathcal{G}$.

Lemma 228. *The* LMA *holds in* $\mathcal{G}'$.

(In [10], LMA(ii) is the definition of segment congruence and LMA(iii) is Proposition 3.1.6.)

The triangle congruence axiom SSS will require more attention because the notion of *congruence* is defined very differently in $\mathcal{G}$ and $\mathcal{G}'$. In [10] a pair of segments are defined to be **congruent** in $\mathcal{G}$ if they have the same length measure (page 51) and a pair of angles are **congruent** if they have the same angle measure (page 100). A pair of triangles are **congruent** in $\mathcal{G}$ if their parts all have the same measure under some correspondence (page 116). In contrast our definition of congruence for $\mathcal{G}'$ applies uniformly to any two figures. They are **congruent** in $\mathcal{G}'$ if some isometry of the entire plane maps one onto the other.

In order to prove that SSS holds in $\mathcal{G}'$, we will need to know that the same triangles are congruent in $\mathcal{G}$ as in $\mathcal{G}'$. To prove that that triangles congruent in $\mathcal{G}$ are congruent in $\mathcal{G}'$ means using the MP-axioms to produce the required isometry from the fact that corresponding parts have the same measure. Lemma 223 will give you a way to do this.

Lemma 229. *Let $\triangle ABC$ and $\triangle DEF$ be triangles in $\mathcal{G}$ and $\mathcal{G}'$. Then $\triangle ABC \cong \triangle DEF$ in $\mathcal{G}'$ if and only if $\triangle ABC \cong \triangle DEF$ in $\mathcal{G}$.*

(Assume $\triangle ABC \cong \triangle DEF$ in $\mathcal{G}$ as shown in Figure 8.1. Again you can use facts in Section 5.1 as well as AAS that is Theorem 6.3.5 of [10]. Assume that CB and FE are maximal length corresponding sides. Show that the feet O and P of the perpendiculars from A and D are between the two base points as illustrated. Now use Lemma 223 to find isometries φ and ψ so that $\chi := \psi^{-1} \circ \varphi$ is an isometry taking $\triangle ABO$ onto $\triangle DEP$. Finally, show that $C^{\chi} = F$.)

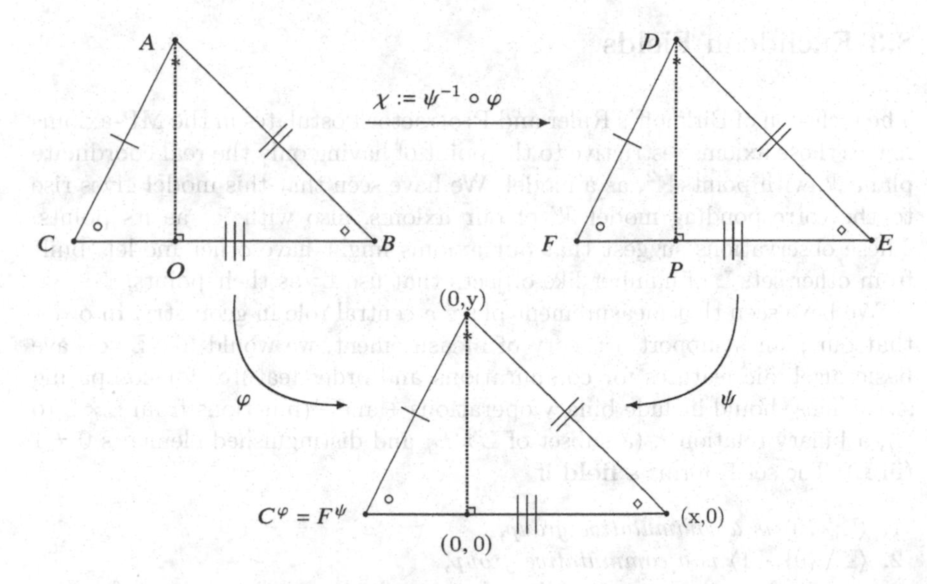

Fig. 8.1 Congruent in $\mathcal{G}$ imples congruent in $\mathcal{G}'$.

As a consequence of Lemma 229, the two definitions of *segment congruence* are also equivalent.

Lemma 230. *Two segments in* **P** *are congruent in $\mathcal{G}$ if and only if they are congruent in $\mathcal{G}'$.*

Lemma 231. *The* SSS *axiom holds in $\mathcal{G}'$.*

(The SSS axiom is Theorem 6.2.3 of [10].)

Theorem 232. *If $\mathcal{G}$ is a model of the MP-axioms, then $\mathcal{G}'$ is a model of our axioms.*

Theorem 233. *The real coordinate plane $\mathcal{R}'$ is a model of our axioms, which are therefore consistent.*

Our proof that the real coordinate plane is a model of our axioms was expedited by borrowing the work of Millman and Parker [10] and by the simplicity of the axioms we used. The initial writing of this book was motivated in part by the goal of demonstrating the consistency of the much more complex and previously unvetted axioms used in [2]. Demonstrating that directly would have been a significantly more difficult task. In Appendix B.10 we will fulfill that goal by drawing on Theorem 233 and on the close connection between [2] and this book.

8.3 Euclidean Fields

The inclusion of Birkhoff's Ruler and Protractor Postulates in the MP-axioms makes those axioms restrictive to the point of having only the real coordinate plane $\mathcal{R}$, with points $\mathbb{R}^2$, as a model. We have seen that this model gives rise to the corresponding model $\mathcal{R}'$ of our axioms, also with $\mathbb{R}^2$ as its points. These observations suggest that our axioms might have other models built from other sets $\mathbb{L}$ of number like objects that use $\mathbb{L}^2$ as their points.

We have seen that measurement plays a central role in geometry. In order that our planes support a theory of measurement, we would like $\mathbb{L}$ to have basic algebraic features for computations and order features for comparing sizes. This should include binary operations $+$ and $\cdot$ (functions from $\mathbb{L}\times\mathbb{L}$ to $\mathbb{L}$), a binary relation $<$ (a subset of $\mathbb{L}\times\mathbb{L}$) and distinguished elements $0 \neq 1$ (in $\mathbb{L}$). The set $\mathbb{L}$ forms a **field** if

1. $\langle \mathbb{L}; +, 0\rangle$ *is a commutative group,*
2. $\langle \mathbb{L} \setminus \{0\}; \cdot, 1\rangle$ *is a commutative group,*
3. $x \cdot (y + z) = x \cdot y + x \cdot z$ *for all* $x, y, z \in \mathbb{L}$.

A field $\mathbb{L}$ is an **ordered field** if

4. $\langle \mathbb{L}; <\rangle$ *is a linear ordering*

and, for all $x, y, z \in \mathbb{L}$ with $x < y$,

5. $x + z < y + z$,
6. $x \cdot z < y \cdot z$ if $z > 0$ *and* $x \cdot z > y \cdot z$ *if* $z < 0$.

Properties (1) through (6) imply that the number 1 has infinite order in the group $\langle \mathbb{L}; +\rangle$ and therefore generates a copy $\mathbb{Z}$ of the integers within $\mathbb{L}$.

Recall from Chapter 3 that the Archimedean Property of the real numbers played a critical role in the proof Lemma 66(iii), which was used to prove the Endomorphism Lemma 67 and the Scaling Theorem 78. They led to the Similar Triangles Theorem 88 which was the key to the Pythagorean Theorem 92 and to defining area measure in Chapter 4. For these reasons we will look for models of our axioms that are coordinate planes of ordered fields with the Archimedean Property. Thus we say that the ordered field $\mathbb{L}$ forms a **number line** (an **Archimedean ordered field**) if

7. *for every positive* $x \in \mathbb{L}$ *there is an integer* $n \in \mathbb{Z}^+$ *such that the rational number* $\frac{1}{n} \in \mathbb{Q}^+$ *is less than* x.

Lemma 234. *An ordered field* $\mathbb{L}$ *is Archimedean if and only if, for every* $x \in \mathbb{L}$*, there is an integer* n *such that* $n - 1 \leq x < n$.

We will review some standard facts about number lines that can be found, for example, in the text [3], *The Number Line through Guided Inquiry* by Clark and Xiao. Since every number line contains $\mathbb{Z}$ and is closed under division, every number line contains the set $\mathbb{Q}$ of rational numbers as well. Thus the **rational number line** is the unique smallest number line.

Using decimal notation we can give a uniform description of the numbers in a number line. An **infinite decimal** is any formal expression

$$\pm n.d_1d_2d_3d_4\ldots$$

where $\pm$ is either $+$ or $-$, where n is a non-negative integer and $d_1, d_2, d_3, d_4, \ldots$ is a sequence of digits that does not end in repeating nines, with the sole exception of $-0.0000\ldots$ that is not an infinite decimal. If $\mathbb{L}$ is a number line and the sequence $\pm n, \pm n.d_1, \pm n.d_1d_2, \ldots$ converges to a number $x \in \mathbb{L}$, then we can use the infinite decimal $\pm n.d_1d_2d_3d_4\ldots$ as a representation of the number x. Theorem 56 of [3] says that every number in a number line $\mathbb{L}$ has a unique infinite decimal representation.

In this context another property of number lines becomes relevant. We say that a number line $\mathbb{L}$ is **complete** if

8. *every non-empty subset of* $\mathbb{L}$ *that has an upper bound has a least upper bound.*

It is not at all clear in advance that there really is a number line that is complete. If a number line $\mathbb{L}$ is complete, then *every* infinite decimal must converge in $\mathbb{L}$. Thus $\mathbb{L}$ would be represented by the set of all infinite decimals. In Chapter 8 of [3] it is shown that $+, -, <, 0$ and 1 can indeed be defined on the set $\mathbb{R}$ of all infinite decimals to form a complete number line. We call this the **real number line**. As a result every number line can be viewed as a subfield of the real numbers. For example, the rational number line $\mathbb{Q}$ consists of all infinite decimals that repeat in finite blocks from some point onward ([3], Theorems 41 and 43).

The natural candidate for another model of our axioms beside the coordinate plane $\mathcal{R}'$ would be the rational coordinate plane $\mathcal{Q}$ whose points are ordered pairs of rational numbers. Lines, circles and rays of $\mathcal{Q}$ could be defined as intersections of lines, circles and rays of $\mathcal{R}$ with $\mathbb{Q}^2$. Unfortunately this idea fails at the very beginning.

Lemma 235. *The Ray Circle Axiom* (RCA) *does not hold in the rational coordinate plane* $\mathcal{Q}$.

Drawing on this experience with Lemma 235, we will add one final restriction on number lines to eliminate those that appear to fail in this way. We say that a number line $\mathbb{L}$ is a **Euclidean field** if

9. *For each positive number* $x \in \mathbb{L}$, *there is a number* $y \in \mathbb{L}$ *such that* $y^2 = x$.

We mention here three familiar and extensively studied Euclidean fields. The completeness property (8) implies that every positive real number has a square root ([3], Theorem 90). Consequently the set $\mathbb{R}$ of all **real numbers** forms a Euclidean field. In Section 8.6 we will see that the field $\mathbb{C}$ of **constructible numbers** forms a Euclidean field. As a third example, a real number is **algebraic** if it is a root of some polynomial over the integers. The theory of field extensions tells us that the set $\mathbb{A}$ of all algebraic numbers forms a subfield of $\mathbb{R}$. If $a \in \mathbb{A}$ is a positive root of a polynomial $p(x)$, then $\sqrt{a}$ is a root of $p(x^2)$ and is therefore in $\mathbb{A}$. Thus the field $\mathbb{A}$ of **algebraic numbers** is a Euclidean field.

It turns out that there are many more Euclidean fields. It follows from Theorem 210 of [3], and is easy to verify, that a subset of $\mathbb{R}$ is a Euclidean field if and only if it contains 1 and is closed under under $+, \cdot, -, ^{-1}$ and $\sqrt{}$. If $\mathbb{S}$ is any subset of $\mathbb{R}$, let $\mathbb{E}(\mathbb{S})$ denote the set of all numbers in $\mathbb{R}$ obtainable from $\mathbb{S} \cup \{1\}$ by a finite number of applications of $+, \cdot, -, ^{-1}$ and $\sqrt{}$. Then $\mathbb{E}(\mathbb{S})$ is itself closed under these operations and contains 1, and is therefore a Euclidean field. Note that $\mathbb{E}(\mathbb{S})$ is countably infinite if $\mathbb{S}$ is countably infinite. It follows that there is an abundance of Euclidean fields, even an abundance of *countably infinite* Euclidean fields. You can prove this by using the first uncountably infinite ordinal $\aleph_1$, which is the smallest uncountable cardinal.

Theorem 236. *There is an uncountable sequence* $\langle \mathbb{L}_\gamma \mid \gamma < \aleph_1 \rangle$ *of countably infinite Euclidean subfields of* $\mathbb{R}$, *where* $\mathbb{L}_\gamma$ *is properly contained in* $\mathbb{L}_\delta$ *if* $\gamma < \delta < \aleph_1$.

The unique smallest Euclidean field, $\mathbb{E}(\varnothing)$, is the topic of our final Section 8.6.

8.4 Models from Euclidean Fields

The goal of this section is to show that the restrictions we have imposed on $\mathbb{L}$ to make it a Euclidean field are sufficient to guarantee that its coordinate plane will indeed be a model of our axioms. As you go through this section you will see how each of those restrictions is called upon to prove one or more of our axioms. It will follow from Theorem 236 that this will give us an abundance of models. Our starting point is the real coordinate plane

$$\mathcal{R}' := \langle \mathbb{R}^2; d^{\mathbb{R}} \rangle,$$

which is a model of our axioms by Theorem 233. For each Euclidean field $\mathbb{L} \subseteq \mathbb{R}$, we will denote by

$$C(\mathbb{L}) := \langle \mathbb{L}^2; d^{\mathbb{L}} \rangle$$

the coordinate plane of $\mathbb{L}$. Points of $C(\mathbb{L})$ are the ordered pairs (x, y) with $x, y \in \mathbb{L}$. The distance function $d^{\mathbb{L}}$ is the restriction of $d^{\mathbb{R}}$ to $\mathbb{L}^2$. All other constructs of $C(\mathbb{L})$ are defined from $d^{\mathbb{L}}$ using the definitions given in Chapters 1 through 7. As an example, $C(\mathbb{R})$ is $\mathbb{R}'$. Our task is to show that $C(\mathbb{L})$ is also a model of our axioms.

We will first show that each of the basic defined constructs of $C(\mathbb{L})$ is an appropriate restriction of a corresponding construct of $C(\mathbb{R})$. We say that a segment of $C(\mathbb{R})$ is an **$\mathbb{L}$-segment** if it is AB for some $A, B \in \mathbb{L}^2$. A ray of $C(\mathbb{R})$ is an **$\mathbb{L}$-ray** if it is $\overrightarrow{AB}$ for some $A, B \in \mathbb{L}^2$. An angle of $C(\mathbb{R})$ is an **$\mathbb{L}$-angle** if it is the union of two $\mathbb{L}$-rays. A line of $C(\mathbb{R})$ is an **$\mathbb{L}$-line** if it is $\overleftrightarrow{AB}$ for some $A, B \in \mathbb{L}^2$. A triangle of $C(\mathbb{R})$ is an **$\mathbb{L}$-triangle** if its vertices are all in $\mathbb{L}^2$. A circle of $C(\mathbb{R})$ is an **$\mathbb{L}$-circle** if its center is in $\mathbb{L}^2$ and its radius is in $\mathbb{L}$. An isometry of $C(\mathbb{R})$ is an **$\mathbb{L}$-isometry** if it takes $\mathbb{L}^2$ onto $\mathbb{L}^2$.

If AB is an $\mathbb{L}$-segment in $C(\mathbb{R})$, then $AB^{\mathbb{L}}$ will denote the segment in $C(\mathbb{L})$ with endpoints A and B. Similarly, if $\mathbf{U}$ is an $\mathbb{L}$-ray, an $\mathbb{L}$-angle, an $\mathbb{L}$-line, an $\mathbb{L}$-triangle, an $\mathbb{L}$-circle or an $\mathbb{L}$-isometry, $\mathbf{U}^{\mathbb{L}}$ will denote the corresponding construct in $C(\mathbb{L})$.

Lemma 237. *Let $\mathbb{L}$ be a Euclidean field. If $\mathbf{U}$ is an $\mathbb{L}$-segment, an $\mathbb{L}$-ray, an $\mathbb{L}$-angle, an $\mathbb{L}$-line, an $\mathbb{L}$-circle or an $\mathbb{L}$-isometry, then $\mathbf{U}^{\mathbb{L}} = \mathbf{U} \cap \mathbb{L}^2$. Conversely, for every segment, ray, line, angle, triangle or circle $\mathbf{X}$ in $C(\mathbb{L})$, there is a construct $\mathbf{Y}$ of the same type in $C(\mathbb{R})$ for which $\mathbf{Y}^{\mathbb{L}} = \mathbf{X}$.*

Note that we have not included isometries in the converse. It is true that every isometry of $C(\mathbb{L})$ is the restriction of an isometry of $C(\mathbb{R})$. But the proof is much more involved and this fact is not required for what we need to do here. Nearly half of our ten axioms can be seen to hold in $C(\mathbb{L})$ from just these observations.

Corollary 238. *Each of the following axioms holds in $C(\mathbb{L})$*: 2PA, 3PA, UNQ, INF, LMA(i,iii).

In general we need to use the fact that an axiom holds in $C(\mathbb{R})$ to prove that it holds in $C(\mathbb{L})$. The axioms PSA, RCA, CCA and EUC all concern the existence of points of intersection of lines and/or circles. Consequently all four will follow immediately from the next theorem. As you prove it, note that all three parts make use of the *field properties* of $\mathbb{L}$ and that (ii) and (iii) both require $\mathbb{L}$ to be a *Euclidean field.*

Theorem 239. *Let $\mathbb{L}$ be a Euclidean field and let $\mathbf{X}$ and $\mathbf{Y}$ be subsets of $C(\mathbb{L})$ that are either*

(i) *two $\mathbb{L}$-lines,*
(ii) *an $\mathbb{L}$-line and an $\mathbb{L}$-circle or*
(iii) *two $\mathbb{L}$-circles.*

Then the points of intersection of $\mathbf{X}$ *and* $\mathbf{Y}$ *are all points of* $\mathbb{L}^2$ *and are therefore the same as the points of intersection of* $\mathbf{X}^{\mathbb{L}}$ *and* $\mathbf{Y}^{\mathbb{L}}$.

Corollary 240. *Each of the following axioms holds in* $C(\mathbb{L})$: PSA, RCA, CCA, EUC.

The remaining axioms, LMA(ii) and SSS, both involve *congruence*, defined in terms of *isometries.* Isometries of $C(\mathbb{R})$ correspond to isometries of $C(\mathbb{L})$ provided that they are $\mathbb{L}$-isometries. What we need here is a convenient way to identify $\mathbb{L}$-isometries. Recall that an $\mathbb{L}$-isometry is an isometry of $C(\mathbb{R})$ that takes *every* point of $\mathbb{L}^2$ to a point of $\mathbb{L}^2$.

Lemma 241. *If an isometry of* $C(\mathbb{R})$ *takes two points of* $\mathbb{L}^2$ *to two points of* $\mathbb{L}^2$, *then it takes every point of* $\mathbb{L}^2$ *to a point of* $\mathbb{L}^2$.

Corollary 242. *Both* LMA(ii) *and* SSS *hold in* $C(\mathbb{L})$.

We can now prove the main theorem of this section. If $\mathbb{L}$ is a field, we denote by $\mathbb{L}^+$ the set of positive numbers in $\mathbb{L}$. If $\mathcal{P}$ is a plane, we will denote by $\mathbb{F}^+_{\mathcal{P}}$ the set of lengths of segments in $\mathcal{P}$.

Theorem 243. *For every Euclidean field* $\mathbb{L}$, *the plane*

$$C(\mathbb{L}) := \langle \mathbb{L}^2; d^{\mathbb{L}} \rangle$$

is a model of our axioms with $\mathbb{F}^+_{C(\mathbb{L})} = \mathbb{L}^+$.

8.5 All Models

In Section 8.2 we answered the first question posed in the introduction to this chapter by showing that $C(\mathbb{R})$ is a model of our axioms, which are therefore consistent. In Section 8.4 we gave a partial answer to the second question,

Exactly which planes are models of our axioms?

by showing that the coordinate planes of all Euclidean fields are models.

We will now complete this answer by showing that coordinate planes of Euclidean fields are the only models. In our context planes have only the distance function d as a primitive construct. Consider planes

$$\mathcal{P} = \langle \mathbf{P}; d \rangle \text{ and } \mathcal{P}' = \langle \mathbf{P}'; d' \rangle.$$

Extending our prior terminology, we define an **isometry** from $\mathcal{P}$ to $\mathcal{P}'$ to be a bijection $\alpha : \mathbf{P} \to \mathbf{P}'$ that preserves distance, that is,

$$d(A, B) = d'(A^\alpha, B^\alpha) \text{ for all } A, B \in \mathbf{P}.$$

Thus, in our geometry, an isomorphism is just an isometry. Planes $\mathcal{P}$ and $\mathcal{P}'$ are **isomorphic** if there is an isometry from one to the other.

We would like to identify all of the models of our axioms as we did with the MP-axioms in Section 8.1. This means finding a set $\mathcal{M}$ of models such that no two models in $\mathcal{M}$ are isomorphic to each other and every model is isomorphic to some model in $\mathcal{M}$. We take

$$\mathcal{M} := \{C(\mathbb{L}) \mid \mathbb{L} \text{ is a Euclidean subfield of } \mathbb{R}\}.$$

In contrast to the MP-axioms, the next theorem tells us that this gives us a very large collection of different models.

Theorem 244. *If $\mathbb{K}$ and $\mathbb{L}$ are two different Euclidean subfields of $\mathbb{R}$, then $C(\mathbb{K})$ is not isomorphic to $C(\mathbb{L})$. Consequently $\mathcal{M}$ contains at least $\aleph_1$ pairwise non-isomorphic countably infinite models of our axioms.*

(Suppose $\alpha : \mathbb{K}^2 \to \mathbb{L}^2$ is an isometry. Use Theorem 243 to show that $\mathbb{F}^+_{C(\mathbb{K})} = \mathbb{F}^+_{C(\mathbb{L})}$ and therefore $\mathbb{K} = \mathbb{L}$. Then apply Theorem 236.)

It remains to show that, for every model $\mathcal{P}$ of our axioms, there is a Euclidean field $\mathbb{L}$ such that $\mathcal{P}$ is isomorphic to $C(\mathbb{L})$. Theorem 243 suggests that we choose $\mathbb{L}$ to be the field whose positive elements are $\mathbb{F}^+_{\mathcal{P}}$. Accordingly, for each model $\mathcal{P}$ of our axioms let $-\mathbb{F}^+_{\mathcal{P}}$ denote the negatives of the numbers in $\mathbb{F}^+_{\mathcal{P}}$, and let

$$\mathbb{L}_{\mathcal{P}} := -\mathbb{F}^+_{\mathcal{P}} \cup \{0\} \cup \mathbb{F}^+_{\mathcal{P}}.$$

Lemma 245. *If $\mathcal{P}$ is a model of our axioms, then $\mathbb{L}_{\mathcal{P}}$ is a Euclidean field.*

Theorem 246. *If $\mathcal{P}$ is a model of our axioms, then $\mathcal{P}$ is isomorphic to the coordinate plane $C(\mathbb{L}_{\mathcal{P}})$ of the Euclidean field $\mathbb{L}_{\mathcal{P}}$.*

(Choose any two perpendicular lines ℓ_X and ℓ_Y of $\mathcal{P}$ intersecting at a point O. Let I_1 and I_{-1} be the two points of ℓ_X and let J_1 and J_{-1} be the two points of ℓ_Y, all at distance one from O. For each $z \in \mathbb{F}^+_{\mathcal{P}}$ let I_z, I_{-z}, J_z and J_{-z} be, respectively, the points on $\overrightarrow{OI_1}, \overrightarrow{OI_{-1}}, \overrightarrow{OJ_1}$ and $\overrightarrow{OJ_{-1}}$ at distance z from O. Now define $\varphi : \mathcal{P} \to C(\mathbb{L})$ as follows. For $A \in \mathbf{P}$ let $A^\varphi = (x, y)$ where the line through A parallel to ℓ_Y intersects ℓ_X at the point I_x and the line through A parallel to ℓ_X intersects ℓ_Y at the point J_y. Show that φ is a bijection and that, for all $A, B \in C(\mathbb{L})$, we have $d^{\mathbb{L}}(A^\varphi, B^\varphi) = d(A, B)$.)

Corollary 247. *If $\mathcal{P}$ is model of our axioms that satisfies the Ruler Axiom, then it is isomorphic to the real coordinate plane $C(\mathbb{R})$.*

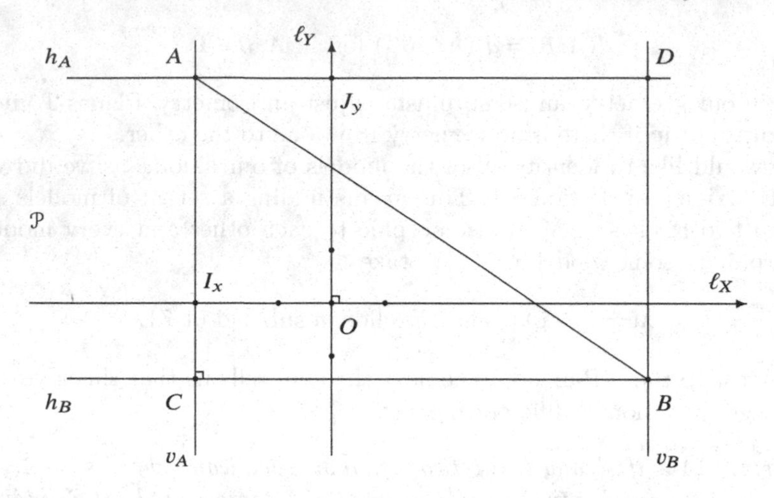

Fig. 8.2 $\varphi : \mathcal{P} \to C(\mathbb{L})$ with $A^\varphi = (x, y)$.

8.6 The Constructible Plane

In this section we will see how valuable and interesting information can be obtained from the coordinate plane of a Euclidean field other than $\mathbb{R}$. We start by imagining a version of the real coordinate plane $\mathbb{R}^2$ in which only the two points (0,0) and (1,0) are initially visible to us. We can make more points and figures visible only by drawing them. We will assume the reader is familiar with basic compass and straightedge constructions as are presented in the first chapter of [2].

We say that a point $(x, y) \in \mathbb{R}^2$ is **constructible** if there is a compass and unmarked straightedge algorithm to draw two lines, a line and a circle or two circles that intersect at (x, y). For example, $(1/2, \sqrt{3}/2)$ is constructible with two applications of the compass. A real number is **constructible** if it is a coordinate of a constructible point. We denote by $\mathbb{C}$ the set of all constructible numbers. It is easy to verify that (x, y) is a constructible point if and only if x and y are both in $\mathbb{C}$. Thus $\mathbb{C}^2$ is exactly the set of constructible points. Recall from Section 8.3 that $\mathbb{E}(\varnothing)$ is the minimal Euclidean field that is contained in every other Euclidean field. We can now give a different description of $\mathbb{E}(\varnothing)$.

Lemma 248. $\mathbb{C} = \mathbb{E}(\varnothing)$.

Taking the distance function $d^{\mathbb{C}} := d^{\mathbb{E}(\varnothing)}$ gives us the alternative designation of the plane $C(\mathbb{E}(\varnothing))$ as

$$C(\mathbb{C}) := \langle \mathbb{C}^2; d^{\mathbb{C}} \rangle.$$

This is known as the **constructible plane**. Its significance was first discovered in the early nineteenth century by a group of French mathematicians, particularly Évariste Galois. The resulting *Galois theory* exposed deep connections between field theory, group theory and geometry. As one of many applications of this theory, it provides important insights as to which numbers are and are not constructible. We will describe two of these applications that are directly applicable to our work.

Since the Protractor Postulate holds in the plane $\mathcal{R}$, the domain $\mathbb{D}$ of the cosine function on $\mathcal{R}$, defined in Chapter 6, is the entire open interval $(0, 90)$. We will be interested in the real number $\cos(20)$. The following theorem of Galois theory is proven using elementary facts about polynomial rings and field extensions, and is well exposited in many texts - for example - see [6], [7] or [11].

Theorem 249. *The number* $\cos(20)$ *is not constructible.*

Outline of Proof. The first step is to use dimensions of field extensions to prove that every constructible number is a root of an irreducible (not properly factorable) polynomial over the rationals whose degree is a power of two, and that every polynomial over the rationals having it as a root is a multiple of that irreducible polynomial.

Next extend Theorem 180 by proving the sum formula for cosine:

$$\cos(x + y) = \cos(x)\cos(y) - \sin(x)\sin(y).$$

Using this and the double angle formulas of Theorem 180 prove the triple angle formula

$$\cos(3x) = 4\cos^3(x) - 3\cos(x).$$

Now apply this with $x = 20$ and simplify it to show that $\cos(20)$ is a root of the polynomial $8x^3 - 6x - 1$. Using properties of the integers, verify that this polynomial has no rational root. Since it is of degree three, it must be irreducible over the rationals. Thus it is not a multiple of a polynomial of degree a power of two so $\cos(20)$ is not constructible. □

We will give two examples of significant applications of this fact.

The early Greeks were well aware of the applications of compass and straightedge constructions as a means of demonstrating the existence of specific figures in the plane. These constructions remain valuable pedagogical tools for beginning students of geometry, and are the topic of Section 1.3 of [2]. A typical construction gives a method to start with an arbitrary angle and produce a ray that bisects it. Aware of this construction, the Greeks raised an obvious question.

Is there a compass and straightedge method to trisect any angle?

After two thousand years of not succeeding in finding such a method, a suspicion arose that there just was no straightedge and compass method to do this. But a demonstration that there was no method seemed equally elusive until the early nineteenth century discovery of Theorem 249.

Theorem 250. *There is no general compass and straightedge algorithm to trisect an arbitrary angle in* $\mathcal{R}$.

The non-constructibility of cos(20) bears directly on Archimedes' method for measuring circles in Chapter 7 of [2] and in our Chapter 7. The technique in both cases is to approximate a circle with n-sided regular polygons. We took n to range over the powers of two starting with four, and demonstrated the existence of regular n-gons for those values of n. In contrast, all values of n greater than two were used in [2]. This leaves open another important question.

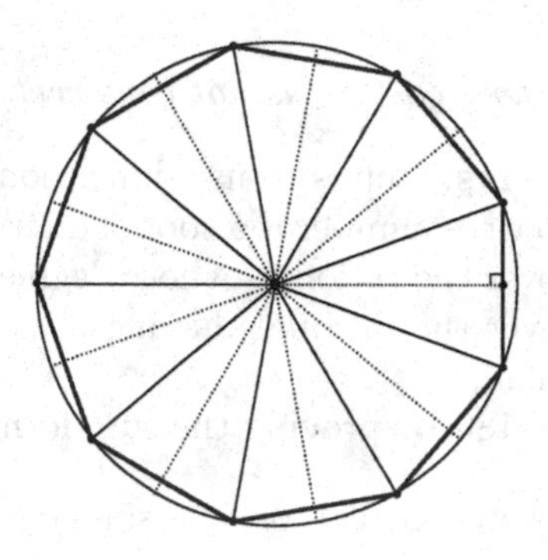

Fig. 8.3 A regular 9-gon.

Do all models of our axioms have a regular n*-gon for every* $n > 2$*?*

You can now answer this question.

Theorem 251. *The constructible plane* $C(\mathbb{C})$ *is a model of our axioms that does not contain a regular 9-gon.*

Appendix A: Axioms

We list here the ten axioms used in this book for a Euclidean plane

$$\mathcal{P} = \langle \mathbf{P}; d \rangle.$$

I. Axioms for MCL#9 Foundational Principles

Non-Triviality Axioms

(2PA) Two Point Axiom (p3). *Given any two points, there is exactly one line containing them.*

(3PA) Three Point Axiom (p3). *There are at least three points that are not collinear.*

Betweenness Axioms

(UNQ) Uniqueness (p4). *Of any three collinear points, exactly one is between the other two.*

(INF) Infinity (p4). If B and D are two points, there there are points A, C and E such that $[ABCDE]$.

(PSA) Plane Separation Axiom (p5). *Let ℓ be a line and let A, B and C be three points not lying on ℓ.*

(i) *If A and B are on the same side of ℓ and B and C are on the same side of ℓ, then A and C are on the same side of ℓ.*

(ii) *If A and B are not on the same side of ℓ and B and C are not on the same side of ℓ, then A and C are on the same side of ℓ.*

Intersection Axioms

(RCA) Ray Circle Axiom (p10). *A circle intersects every ray emanating from a point inside the circle.*

D. M. Clark, S. Pathania, *A Full Axiomatic Development of High School Geometry*,
https://doi.org/10.1007/978-3-031-23525-2

(CCA) Circle Circle Axiom (p11). *If a circle contains a point outside and a point inside another circle, then the two circles intersect in exactly two points which lie on opposite sides of the line containing their centers.*

II. Axioms from [2].

(LMA) Length Measure Axiom (p9). *The length measure function $\mathcal{L}$ has the following properties.*

(i) *There are two points O and I such that $\mathcal{L}(OI) = 1$.*
(ii) *Segments are congruent if and only if they have the same length.*
(iii) *$\mathcal{L}(AC) < \mathcal{L}(AB) + \mathcal{L}(BC)$ for any three non-collinear points A, B and C (the Triangle Inequality).*

(SSS) Side-Side-Side (p11). *If $\triangle ABC$ and $\triangle XYZ$ are triangles with $AB \cong XY$, $AC \cong XZ$ and $BC \cong YZ$, then $\triangle ABC \cong \triangle XYZ$.*

(EUC) Euclidean Parallel Axiom (p25). *For every line ℓ and every point P not on ℓ, there is at most one line containing P that is parallel to ℓ.*

Appendix B: MCL#9

This appendix is provided for the benefit of users of the first author's prior text [2]. Our preface describes the goals and methods of that book, which covers the standard topics of high school geometry in a single undergraduate guided inquiry course for teachers. It gives those teachers a solid working understanding of what it means to establish a fact from given information and a solid working knowledge of the topics they themselves will later need to teach. However, achieving all of this in a single semester with students having only a high school background in geometry required some compromises. In this appendix we will give an outline of the conscious omissions and logical gaps put into [2] and how they have been filled in this text.

1. Foundational Principles. The word *geo-metry* comes from the Greek words for *earth* and *measure* and refers to the science of measurement of physical space. From this point of view consider the following two statements.

1. *The area of a circle of radius r is between* $3.14158r^2$ *and* $3.14160r^2$.
2. *A line containing a point inside a circle intersects the circle.*

These statements are both true in the real Euclidean coordinate plane. But they differ in a significant way. The first would never have been discovered without some very careful analysis of circles. In contrast, the second would immediately be taken as a fundamental fact about lines and circles on a flat surface. Eighteenth century geometers were perplexed by this distinction, not seeing any clear means to determine which statements required a proof and which statements were obviously true of physical space.

Nevertheless, beginning students of geometry were quite willing to recognize and respect this distinction without having any clear line drawn between the two. As a result, Euclid's geometry inspired a broad range of students and scholars for well over two millennia. The goal of [2] was to give a formulation of geometry that offered the same benefits of Euclid, but this time with strategically chosen omissions together with the insights offered by modern mathematics. The first page of the first chapter begins with a seemingly innocuous statement and definition:

D. M. Clark, S. Pathania, *A Full Axiomatic Development of High School Geometry*,
https://doi.org/10.1007/978-3-031-23525-2

In plane geometry we study figures that can be drawn on a flat surface, like a sheet of paper. We say that a figure **X** *is* **congruent** *to a figure* **Y** *(written* $\mathbf{X} \cong \mathbf{Y}$*) if we can orient one, or a copy of one, on top of the other so that they match exactly.*

Clearly these are statements about physical space and figures in physical space. In this informal setting beginning students learn to make reasonably compelling arguments based on the imprecise information they have. After Chapter 1 they transition to proving theorems using precise definitions and axioms. Yet the initial context opens the opportunity to draw on what they know about physical space at carefully chosen points along the way.

The unstated assumptions used by Euclid and [2] primarily fall into three classes of simple statements whose validity in physical space would rarely be questioned. We list them here with an example of each.

Non-triviality. *For every line, there is a point not on that line.*
Betweenness. *Of three points on a line, one is between the other two.*
Intersections. *A ray from a point within a circle intersects the circle.*

The preface of [2] labels these kinds of statements collectively as *foundational principles* that students should be allowed to freely use without proof just as students of Euclid had always done. Experience with [2] has shown this strategy to work very well with the intended audience. But it did mean that [2] fell far short of giving a full axiomatic development of high school geometry. In particular the betweenness relation [...] is technically used as a primitive construct, along with length, area and angle measure, although no governing axioms for [...] are given. Thus a [2]-plane has four primitive constructs: $\mathcal{P} = \langle \mathbf{P}; [...], \mathcal{L}, \mathcal{A}, \mathcal{D} \rangle$.

Solution: The axioms given in our Chapter 1 include the LMA and SSS from [2] and replace the Foundational Principles of [2] with seven Hilbert style axioms about non-triviality, betweenness and intersections. It then gives proofs of all of the foundational principles that will be needed later. One additional axiom, EUC from [2], is given in Chapter 3. All other facts in Chapters 1 to 7 can be proven from these axioms without any compromises.

2. Congruence. The notion of congruence is defined transformationally in Chapter 1 of [2] in terms of isometries. But isometries are only defined informally by referring to physical objects and physical actions. Later the SSS and SAS axioms are given for triangle congruence. But no means is given of proving congruence of figures other than triangles and their associated parts. Yet it is intuitively clear, for example, that two circles with the same radius should be congruent.

Solution: We say that figures are congruent if some isometry of the entire plane takes one onto the other. This makes the notion of congruence applicable to all figures rather than just a few isolated types. For example, it allows us to prove statements like "*Two circles are congruent if and only if they* *have the same radius.*" (Theorem 17) and "*Two rectangles are congruent if and only if their corresponding sides are congruent.*" (Theorem 64) that do not follow from the axioms of [2].

3. Real Numbers. College bound children normally enter high school with a good working knowledge of the rational number system. In high school they gradually learn about more numbers that are described in terms of a defining property rather than a numerical value, e.g.,

$$\sqrt{14},\ \pi,\ \frac{\sqrt{14}}{\pi},\ \log_{10}(23),\ \sin(1.3),\ e, \ldots$$

It may or may not be mentioned that most of these numbers are not rational; they are just "numbers" that can be approximated by rational numbers. They can be thought of as collectively forming a continuous whole called "the real number line". But no explanation is given as to what a real number is, much less how to do arithmetic operations with real numbers.

They are unlikely to ever learn more about the real number system beyond this in either high school or college unless they major in mathematics. Even then, teachers may or may not have a course like [3] in which they will become familiar with the real number system. And even those who do are unlikely to take such a course before they take a course like [2] in geometry.

It was for this reason that Birkhoff's Ruler and Protractor postulates were not mentioned in [2]. Specifically, the Length Measure Axiom (LMA) of [2] asserts the existence of a length measure function $\mathcal{L}$ that assigns every segment AB to a positive number $\mathcal{L}(AB)$. But it intentionally refrains from invoking Birkhoff's Ruler Postulate to conclude that every positive real number is the length of some segment because the intended readers would have no idea what that meant.

This left unanswered the question as to what the set $\mathbb{F}^+_{\mathcal{P}}$ of all lengths of segments in a plane $\mathcal{P}$ could be. Is it necessarily the set $\mathbb{R}^+$ of all positive real numbers? If not, what could it be?

Solution: Both questions are fully answered in Chapter 8. Theorem 246 says that, for any model $\mathcal{P}$ of our axioms, the set $\mathbb{F}^+_{\mathcal{P}}$ is the set of all positive numbers in some Euclidean subfield $\mathbb{L}_{\mathcal{P}}$ of the reals and that $\mathcal{P}$ is isomorphic to the coordinate plane of $\mathbb{L}_{\mathcal{P}}$. Conversely, Theorem 243 says that the coordinate plane of every Euclidean field $\mathcal{L}$ is a model of our axioms in which the positive numbers in $\mathcal{L}$ are exactly the lengths of its segments.

Readers interested in Birkhoff's geometry do have the option to assume the Ruler Postulate and read "$\mathbb{F}^+$" as "$\mathbb{R}^+$" throughout. This will simplify or eliminate many parts of this text, particularly in Chapter 8.

4. Closed Regions. Chapter 3 of [2] on area measure begins with a brief discussion of the kinds of figures that we will consider to have an area. These figures are called *closed regions*, but that phrase is only explained by a drawing of several examples. The measure of areas depends on the Area Measure

Axiom 6 of [2] that states properties of areas of closed regions. To legitimately apply this axiom to a figure would require first proving that the figure is a closed region, and that is not possible without an exact definition of *closed region.*

In order to give a full mathematical development, it would also be necessary to define the measure function $\mathcal{A}$ and then prove that it has the properties listed in Axiom 6 of [2].

Solution: Filling in these gaps has required not only adding further content, but also required some reorganization of the content of [2]. We have done this in two stages. In Chapter 4 we identify a limited class of closed regions, called *polygonal regions*, define an area measure function $\mathcal{A}$ on those regions and prove our Theorem 120 saying that it satisfies the first three properties listed in Axiom 6 of [2]. In our Chapter 7 we build on these results by defining a broader class of closed regions that includes both circular discs and polygonal regions and an area function $\mathcal{J}$ that is an extension of $\mathcal{A}$ to that class of regions. Doing this requires us to prove Theorem 199 saying that $\mathcal{A}$ satisfies the fourth property listed in Axiom 6 as well.

Our definition of the area function $\mathcal{A}$ is built from the definition of the area of a triangle with base b and height h. The fact that all three choices of b and h give the same value for $bh/2$ is an immediate consequence of the Similar Triangles Theorem of [2], our Theorem 88, which is the culmination of Chapter 5 in [2]. In order to apply this theorem to area measure, we have changed the order of Chapters 3, 4, 5, giving us now first Chapter 3: Similar Figures, then Chapter 4: Area Measure and then Chapter 5: Angle Measure.

5. Rectangle Area Theorem. Area formulas for familiar polygons were established in [2] from the rule for computing the area of a rectangle. This rule was given as the Rectangle Area Theorem. Students were asked to prove it in a very special case, and then use it in general without proof. Omitting this proof saved considerable time but left a significant gap that needed to be filled.

Solution: The Rectangle Area Theorem is our Theorem 123.

6. Pythagorean Theorem. A classical proof of this theorem is given in [2] that makes essential use of both area measure and angle measure. However, this proof comes before the similarity chapter in [2] while our definition of area begins with a critical application of the Similar Triangles Theorem 88 in Lemma 102. Consequently we can not justify the proof in its location in [2].

Solution: We have resolved this problem by putting that proof after the similarity, area measure and angle measure chapters as Theorem 175 since it requires all three as background. Doing this led us to add Theorems 52, 92, 101 and 126 showing four other proofs and variations of this theorem, each drawing on newly acquired background information.

7. Dilations. Like the definition of congruence, [2] used a transformational definition of similarity based on the notion of a dilation. The *dilation* with center O and scaling factor $x \in \mathbb{R}^+$ was defined to be the transformation on the set of all points taking O to itself and each point $P \neq O$ to the point P' on ray $\overrightarrow{OP}$ whose distance from O is x times the distance from O to P.

This definition immediately begs the question (not raised in [2]) as to whether there really is a point $P' \in \overrightarrow{OP}$ at that distance from O. Birkhoff's Ruler Postulate guarantees that such a point P exists for every positive real number x. Since we do not use the Ruler Postulate, we must ask: For which positive real numbers x are we guaranteed the existence of the required point P' for each choice of O and P?

Solution: Assume that O is a point and x is a positive real number such that, for every point $P \neq O$, there is a point $P' \in \overrightarrow{OP}$ with $\mathcal{L}(OP') = x\mathcal{L}(OP)$. If we take OP to be a unit interval, then the length of OP' would be $x1 = x$. Thus $x \in \mathbb{F}^+$. Conversely, we used the Endomorphism Lemma 67 to prove the Scaling Lemma 69 saying that, for every $x \in \mathbb{F}^+$, every point O and every point $P \neq O$, there does exist a point P' on $\overrightarrow{OP}$ such that the length of OP' is x times the length of OP. Thus the scaling factors for dilations are exactly the numbers in $\mathbb{F}^+$.

8. Angle Measure. Chapter 6 of [2] presents the trigonometry of acute angles and uses it to find missing measurements in an assortment of partially specified triangles. If the length of a side or the degree measure of an angle is specified, it is necessary to ask if there really is a segment or angle with that measurement in the plane. This issue is not mentioned in [2]. It does not directly concern the development in this book, but is relevant to the context that this book presents.

Solution: On page 71 of Chapter 6 we have addressed this issue. There we explain how Birkhoff's Ruler and Protractor Postulates will guarantee the presence of these segments and angles by restricting the models of the axioms to only the real number coordinate plane. There we also prove Theorem 178 saying that, in our context, the Ruler Postulate alone is sufficient since it implies the Protractor Postulate.

In Chapter 8 we saw that our weaker axioms embrace many more models. Theorem 244 implies that there are many countably infinite models of our axioms in which the set of angle measures is countably infinite and therefore not all of the open interval (0,180). Theorem 249 gives a specific model which does not contain an angle with degree measure 20.

9. Circles. Chapter 7 of [2] used Archimedes method of approximating circles with regular circumscribed polygons to find the circumference and area of a circle. This required establishing four assertions.

(i) The ratio of the circumference to the diameter of any two circles is the same.
(ii) The ratio of the area to the square of the radius of any two circles is the same.
(iii) Those two ratios are the same number, denoted by the Greek letter π. The circumference c and the area a of any circle of radius r are given by $c = 2\pi r$ and $a = \pi r^2$.
(iv) The value of π is given by $\pi = \lim_{n\to\infty} n \tan\left(\frac{90}{n}\right)$.

Compelling arguments using diagrams and experimental calculations were given in [2] for each of these. Although they are intuitively appealing adaptations of Archimedes work, none of these arguments rise to the level of a proof since several items are missing:

1. a demonstration that the required polygons actually exist,
2. a definition of circumference for (i) and (iii),
3. a definition of circle area for (ii) and (iii) and
4. a definition of π for (iii) and (iv).

Solution: In our Chapter 7 we fill in each of these four gaps and give complete proofs of (i), (ii), (iii), (iv) as follows.

1. For each positive integer n we define the inscribed convex 2^{n+1}-gon $\mathbf{I}_n$ in Section 7.2 and the circumscribed convex 2^{n+1}-gon $\mathbf{O}_n$ in Section 7.3.
2. The *circumference* of a circle is defined on page 98.
3. The *area* of a circle is defined on page 89 as the Jordan measure of the corresponding closed disk, provided that closed disk is Jordan measurable.
4. The number π is defined in Theorem 212.

Theorems 219 and 220 together show that the ratios in (i) and (ii) are both π by establishing the formulas of (iii). Item (iv) is Theorem 212.

10. Consistency. Do the axioms used in [2] lead to a consistent theory of geometry, that is, do they have a model?

Solution: In Chapter 8 we answered this question affirmatively for our axioms. We will now use the results of Chapter 8 to show that the answer is also affirmative for the [2]-axioms. We do this by showing that each model $\mathcal{P}$ of our axioms, as described in Section 8.5, leads to a model $\mathcal{P}^*$ of the [2]-axioms with the same set of points. To do this we start with any model $\mathcal{P} = \langle \mathbf{P}; d \rangle$ of our axioms. Using $\mathcal{P}$ we take

$$\mathcal{P}^* = \langle \mathbf{P}; [\ldots], \mathcal{L}, \mathcal{A}, \mathcal{D} \rangle$$

to be the [2]-plane with its four primitive constructs defined as they are in $\mathcal{P}$ starting with the distance function d. Appendix E gives page numbers for each of these definitions.

We would like to show that $\mathcal{P}^*$ is a model of the [2]-axioms, including the three unproven theorems used in [2] as further axioms. In order to do that we first need to recognize that [2] falls short of offering a proper axiomatic theory because it uses five terms or phrases that are not carefully specified:

isometry, congruent figures, closed region, scaling factor, dilation.

Congruence is defined in [2] in terms of physical actions as we quote in item 1. Foundational Principles above. Isometries are those physical actions (page 4). Closed regions, whose areas are to be found, are defined by drawings of several examples (page 29). Dilations are defined by scaling factors (page 51), but nothing is said about which positive real numbers are scaling factors. In order to give a valid interpretation to the axioms and theorems of [2], we recast each of these terms as a defined notion as they are in this text.

1. An **isometry** is a distance preserving bijection.
2. Figures **X** and **Y** to **congruent** if some isometry takes one onto the other.
3. A **closed region** is a polygonal region. In Chapter 7 we expand the definition to include all Jordan measurable figures.
4. A **scaling factor** is a number in $\mathbb{F}^+$.
5. A **dilation** for each scaling factor is defined as we have in Chapter 3.

With these adjustments the axioms and theorems of [2] become well formulated mathematical statements that have the same meaning in $\mathcal{P}^*$ as they have in $\mathcal{P}$. Consequently we can demonstrate that the [2]-axioms are true in $\mathcal{P}^*$ by proving them true in $\mathcal{P}$. Appendix D gives a full listing of the axioms and theorems in [2] together with the axiom or theorem of this book that establishes that they are true in $\mathcal{P}$. Since this list includes the axioms of [2], each model $\mathcal{P}$ of our axioms described in Section 8.5 corresponds to a model $\mathcal{P}^*$ of the [2]-axioms with the same set **P** of points. In particular, the axioms of [2] are consistent.

Appendix C: Font Guide

$A, B, C, \ldots, X, Y, Z$
— Capital Italic: **points**

$\mathbf{P}, \mathbf{X}, \mathbf{Y}, \mathbf{Z}$
— Captal Roman Bold: **sets of points (figures)**

$\overline{\mathbf{X}}$ (where $\mathbf{X}$ is a convex polygon or circle)
— Overline Capital Roman Bold: **X together with its inside**

ℓ, m, n, q, r, s, t
— Lower Case Italic: **lines and rays**

a, b, c, x, y, z
— Lower Case Italic: **numbers**

f, g and r
— Lower Case Italic: **functions from $\mathbb{R}$ to $\mathbb{R}$**

$\alpha, \beta, \varphi, \psi, \chi$
— Lower Case Greek: **isometries**

γ, δ
— Lower Case Greek: **infinite ordinals**

$\mathcal{A}, \mathcal{L}, \mathcal{M}$
— Capital Script: **real valued measure functions**

$\mathbb{F}^{+}, \mathbb{F}, \mathbb{K}, \mathbb{N}, \mathbb{R}, \mathbb{Z}$
— Capital Black Board Bold: **numerical structures**

$\mathcal{P}, \mathcal{R}, \mathcal{C}(\mathbb{L})$
— Capital Math Script: **geometric planes**

D. M. Clark, S. Pathania, *A Full Axiomatic Development of High School Geometry*,
https://doi.org/10.1007/978-3-031-23525-2

Appendix D: Theorem Index

Our Chapters 1 through 7 give a full axiomatic development of the topics of high school plane geometry that parallels the more elementary development of those same topics in Chapters 1 through 7 of [2]. This index lists all items from those chapters of [2] that take part in its axiomatic development. Each of these listings is followed by the item in the present text where that result is established, thereby showing that the facts presented in [2] can all be proven from our axioms. Chapter numbers refer to [2].

Ch1. Congruent Figures

Axiom 1(i),(ii) .... LMA, p9
Axiom 1(iii)........Definition, p2
Axiom 2 ....SSS, p11
Problem 20.........Theorem 30
Axiom 3...........Theorem 22

Ch2. Axioms, Thms & Proofs

Theorem 29........Lemma 20
Theorem 30........Corollary 24
Theorem 31........Theorem 26
Theorem 32........Theorem 29
Theorem 33........Theorem 28
Theorem 34........Theorem 27
Theorem 35........Theorem 25
Theorem 36........Theorem 31
Theorem 37........Theorem 32
Corollary 38........Corollary 33
Theorem 39........Theorem 34
Axiom 4...........Theorem 35
Theorem 40........Theorem 37
Corollary 41........Corollary 38
Corollary 42........Corollary 39
Theorem 43........Theorem 40
Theorem 44........Theorem 41
Theorem 45........Theorem 42
Theorem 46........Theorem 43
Theorem 47........Theorem 36
Theorem 48........Theorem 44
Theorem 49........Theorem 45
Corollary 50........Corollary 46
Corollary 51........Corollary 47
Corollary 52........Corollary 48
Theorem 53........Theorem 49
Theorem 54........Theorem 51
Problem 55.........Theorem 50

Ch3. Area Measure

Axiom 5 ....EUC, p25
Corollary 59........Corollary 53
Theorem 60........Theorem 54
Corollary 61........Corollary 55
Corollary 62........Corollary 58
Theorem 63........Theorem 59
Theorem 64........Theorem 60
Theorem 65........Theorem 62

D. M. Clark, S. Pathania, *A Full Axiomatic Development of High School Geometry*,
https://doi.org/10.1007/978-3-031-23525-2

Ch4. Angle Measure

Ch5. Similar Figures

Ch6. Trigonometric Ratios

Ch7. Circle Measure

Appendix E: Notation Index

D. M. Clark, S. Pathania, *A Full Axiomatic Development of High School Geometry*,
https://doi.org/10.1007/978-3-031-23525-2

Bibliography

1. G. D. Birkhoff, "A Set of Postulates for Plane Geometry (Based on Scale and Protractors)", Annals of Mathematics 33 (1932), 329–345.
2. D. Clark, *Euclidean Geometry: A Guided Inquiry Approach*, Math Circles Library #9, American Mathematical Society (Rhode Island) and Mathematical Sciences Research Institute (California) (2012).
3. D. Clark and X. Xiao, *The Number Line through Guided Inquiry*, AMS/MAA Textbooks #69, American Mathematical Society (Rhode Island) (2022).
4. Euclid of Alexandria, *The Thirteen Books of Euclid*, Cambridge University Press (1926). Authorized translation by Thomas Heath.
5. G. Frege, *Basic Laws of Arithmetic* (*Grundgesetze der Arithmetik*), Vol. 2 (1903).
6. J. Gallian, *Contemporary Abstract Algebra*, 8th Edition, Brooks and Cole (2013).
7. R. Hartshorne, *Geometry: Euclid and Beyond*, Springer-Verlag (2000).
8. D. Hilbert, *Grundlagen der Geometrie* (1899). English translation: E. J. Townsend, *The Foundations of Geometry*, Open Court Pub. Co., Illinois (1902).
9. C. Jordan, "Remarques sur les intégrales définies", J. Math. Pures Appl., 8 (1892), pp. 69–99.
10. R. Millman and G. Parker, *Geometry: A Metric Approach with Models*, Springer-Verlag (1981).
11. E. Moise, *Elementary Geometry from an Advanced Standpoint*, 3rd Edition, Addison-Wesley Publishing Company (1990).

D. M. Clark, S. Pathania, *A Full Axiomatic Development of High School Geometry*,
https://doi.org/10.1007/978-3-031-23525-2

Index

D. M. Clark, S. Pathania, *A Full Axiomatic Development of High School Geometry*,
https://doi.org/10.1007/978-3-031-23525-2

Printed in the United States
by Baker & Taylor Publisher Services

Printed in the United States
by Baker & Taylor Publisher Services